Stephan E. Wolf

Nonclassical Crystallization of Bivalent Metal Carbonates

Stephan E. Wolf

Nonclassical Crystallization of Bivalent Metal Carbonates

On nonclassical liquid intermediates, symmetry-breaking phase-selection of metal carbonates, and their morphosynthesis aided by mesocrystallinity

Südwestdeutscher Verlag für Hochschulschriften

Imprint

Any brand names and product names mentioned in this book are subject to trademark, brand or patent protection and are trademarks or registered trademarks of their respective holders. The use of brand names, product names, common names, trade names, product descriptions etc. even without a particular marking in this work is in no way to be construed to mean that such names may be regarded as unrestricted in respect of trademark and brand protection legislation and could thus be used by anyone.

Publisher:
Südwestdeutscher Verlag für Hochschulschriften
is a trademark of
Dodo Books Indian Ocean Ltd., member of the OmniScriptum S.R.L Publishing group
str. A.Russo 15, of. 61, Chisinau-2068, Republic of Moldova Europe
Printed at: see last page
ISBN: 978-3-8381-2539-8

Zugl. / Approved by: Mainz, Johannes-Gutenberg Univ., Diss,. 2009

Copyright © Stephan E. Wolf
Copyright © 2011 Dodo Books Indian Ocean Ltd., member of the OmniScriptum S.R.L Publishing group

For my parents.

All science is a palimpsest.

o5. o4. o7
Inspired by *Nineteen Eighty-Four*.

Hypotheses are nets; only he who casts will catch...

Novalis
Dialogen und Monolog, Dialogen 5., 1789.

Contents

1 Introduction: Nonclassical Crystallization *in vivo et in vitro*   1

2 Research Objectives   11

3 Nonclassical Liquid Intermediates During the Homogenous Formation of Calcium Carbonate   13

4 Nonclassical Homogenous Formation of Divalent Metal Carbonate Minerals   33

5 Stabilization and Destabilization of the Transient Liquid Calcium Carbonate Precursor Phase by Ovo-Proteins   41

6 Cryogenic Molding of Nanotubes in Mesocrystalline Ice   57

7 Nonclassical and Symmetry-Breaking Phase Selection of Calcium Carbonate Triggered by Amino Acids   65

8 Résumé   89

Appendix   93

1 Introduction: Nonclassical Crystallization *in vivo et in vitro*

Nucleation, the emergence of a not yet existing crystalline phase in its metastable mother phase such as a supersaturated solution, is one of the most crucial and fundamental aspects of phase transitions in general. Phase transitions are of enormous relevance in technical, biological, and ecological processes, be they phase separations or precipitation reactions.

Even apparently simple systems like calcium carbonate exhibit complex chemical behavior and are strongly affected by coordination chemistry. Calcium carbonate is not only a model system of academic interest since undesired precipitation of calcium carbonate remains a permanent challenge of all-day life. The branches of industry which have to cope with scale formation and its prevention are omnipresent, *e. g.* oil and gas production, water piping, energy supply or laundry cleaning. The directed and controlled precipitation during calcium carbonate production provides the basis of a legion of industrial applications. Calcium carbonate and the chemical additives involved in its productions are thus megascale products: Calcium carbonate serves as filler for paper (worldwide $\sim$ 12 million tons *p. a.*) and synthetics, dyes or renders (worldwide $\sim$ 19 million tons *p. a.*), construction material in form of cement or burnt lime, flux in steel production or mineral fertilizer in agriculture.[1] In 2007, the world demand of calcium carbonate added up to 85 million tons and showed an annual increase of nearly 9%.[2,3] *Industrial Importance*

Biomineralization, its comprehension and imitation in biomimetic mineralizations can be regarded as the ultimate in phase transitions and may pave the way for complex mesostructured and functional materials.[4] In Nature's concept of mineralized tissues, the mere constituents exhibit only moderate mechanical properties: The mineral components like silica or calcium carbonate are fragile or brittle in nature but of abundant supply whereas the organic constituents are pliable, soft but quite tough. Following the Aristotelian dictum that "the whole is more than the sum of its parts",[5] Nature succeeds in giving its mineralized tissues extraordinary strength and toughness by applying a superior ordered compositional structure which *Biological Relevance* spans over several orders of magnitude.[4,6,7] Ernst Haeckel, natural scientist and philosopher, extensively compiled in his works the intriguingly complex and elegant shape of biominerals, e. g. in his popular "Kunstformen der Natur" released in 1899–1904. Albeit from a philosophical and monistic point of view, he was maybe the first who studied simple imitation of biogenic minerals *in vitro* and devoted his book "Kristallseelen, Studien über das anorganische Leben" to this topic in 1917.[8]

Biomineralization Challenges Classical Theories The formation of calcium carbonate in general has been studied for more than a century and especially its biogenic formation was intensively investigated in the past two decades. Still little is known about the early stages of calcium carbonate formation and of the exact methods Nature applies in order to build up biominerals. In biomineralization, one perpetually observes crystal morphologies and crystal formation processes which challenge the classical concepts of crystallization. Latter are outlined in the appendix (Sec. 8). Biominerals often exhibit an exceptional shape with curved surfaces which does not resemble the crystal habit which would evolve under conditions of equilibrium. Figure 1.1 may serve as an example; both drawings compile the shaping of calcium carbonate and their differences are obvious. The calcium carbonate crystals of Goldschmidt's compilation (Fig. 1.1a) show planar surfaces of high symmetry. Their appearance can be explained on the basis of established rules.[9-12] Biogenic molded calcium carbonte illustrated by Haeckel in Fig. 1.1b is in sharp contrast to them. The two enlarged hook-like spicules consist of calcium carbonate as well as the three-folded spicules in the lower cross sections of the calcisponge *Sycon*. The most prominent and celebrated example of exquisite molding of calcite is presumably *Emiliania huxleyi*, which is a species of coccolithophores. These single-celled algae form a calcareous exoskeleton, which is hierarchically assembled from round shield elements. In turn, these elements consist of 35 complex shaped individual elements (Fig. 1.2a). Each of these units consists of one single calcite crystal although its hammer-headed stirrup-like form is far away from the classical trigonal-rhombohedral calcite shape.[6] Biomimetic approaches are still not able to reproduce the morphogenesis of *Emiliania huxleyi*. In case of the coccoliths *Discosphaera tubifera* (Fig. 1.2b), Mukkamala et al. could roughly mimic the trumpet-shape by applying a derivate of the polycarboxylate ethylenediaminetetraacetic acid as an additive.[13] However, the detailed morphogenesis of the reported microtrumpets composed of nanocrystalline calcite still remains unsolved.

Only the hard tissue of the major taxon *Crustacea* consists of amorphous calcium carbonate (ACC) in analogy to silica skeletons of radiolarians, sponges, plants phytoliths or diatoms.[17] In most other cases, ACC acts as a temporary calcium carbonate storage prior the formation of a hard tissue. Beniash et al. could show ACC as a transient precursor phase prior spicule formation

Biomineralization Challenges Classical Theories

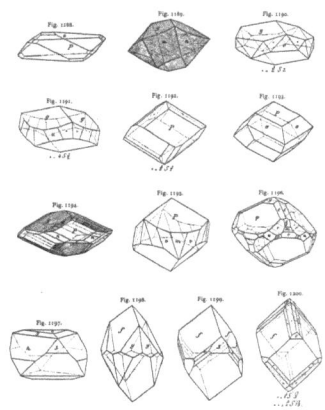 

(a) Examples of calcite crystals as tabulated by Victor Goldschmidt in his "Atlas der Krystallformen" from 1913.[14]

(b) Calcarous sponges (*Calcarea* or *Calcispongiae*) as depicted in Ernst Haeckel's "Kunstformen der Natur" dating back to 1904.[15]

Figure 1.1 Comparison of geological and biogenic shaping of calcium carbonate as reported in the first two decades of the 20th century.

 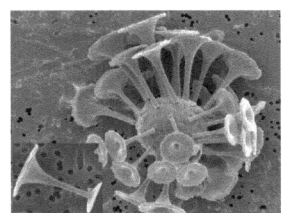

(a) Coccosphere of the calcareous algae *Emiliania huxleyi* surrounded by its coccoliths (proximal view). Scale bar: 1 µm

(b) Coccosphere and one of its coccoliths (inset) of the calcareous algae *Discosphaera tubifera*. Scale bar: 1 µm

Figure 1.2 Two representatives of calcareous algae *Emiliana huxleyi* (*Noelaerhabdaceae*) and *Discosphaera tubifera* (*Rhabdosphaeraceae*). Images taken from Young et al.[16]

in sea urchin larvae and Weiss *et al.* demonstrated this analogously in larvae of molluscan bivalves.[18,19] Calcium carbonate is unstable in its amorphous state and tend strongly to transform into a crystalline phase. Organisms like those mentioned above have thus developed strategies in order to prevent undesired phase transformation and induce directed mineralization to a distinct polymorph, be it calcite or aragonite.[18–22] The stabilization strategies seem to base on biomacromolecules (*e. g.* GlX-, Ser- and AsX-rich proteins, glycoproteins, polysaccharides, or proteoglycans), which are often highly acidic in nature.[23,24] Biogenic ACC contains frequently magnesium and phosphate ions, which may stabilize of the amorphous state.[17,25] Interestingly, biogenic ACC from different species exhibit identical chemical composition but structural differences according to X-ray absorption spectroscopy.[26]

One further striking and long known example of excellent control over mineralization processes is presented us by the abalone shell. It consists of two different calcium carbonate polymorphs, namely calcite in the outer portion of the shell and aragonite in the inner nacreous layer.[27] If soluble proteins extracted from the abalone nacreous layer are incubated with calcite seed crystals in a supersaturated calcium carbonate solution, they induce the nucleation and growth of aragonite needles on the (104) faces of the seed crystals.[28] Before this finding it had been assumed that a complete deposition of a nucleation layer sheet is essential to induce a switch in crystal phase. By atomic force microscopy, changes in the atomic lattice of a calcite seed crystal were directly observed and show the soluble nacre proteins inducing an (001) aragonite growth on a (104) calcite surface.[29] Such a directed switch in phase is still quite complicated to achieve *in vitro* by traditional chemical means.

In short, organism have developed strategies to give biominerals superior properties in order to serve as sensors, skeletal support or protection of soft tissues so that they excel their purely inorganic counterparts. ACC plays a pivotal role in these strategies. Being inspired by this strategy, its concepts were applied in biomimetic material chemistry in numerous bottom-up approaches for synthesis of advanced materials.[30–33] Today, material science is only partially able to produce corresponding materials at identical length scale and quality. Furthermore, biomineralization mechanisms cannot be described within the classical pictures of crystallization and phase separation.

Nonclassical Crystallization Over the last few years, numerous experiments *in vitro* seconded and substantiated as well the principles of a nonclassical crystallization. It can be assumed that the formation of an amorphous precursor is the basis step of nonclassical crystallization. Then, two major different characteristics of nonclassical crystallization can be classified:[34] *(a)* The formation of an intermediate amorphous precursor, be it amorphous or even liquid which later may transform to crystalline material. *(b)* Oriented attachment of nano-sized particles in a self-organized manner to yield complex morphologies. The term 'mesocrystallization' describes the special case of a three-dimensional oriented attachment of colloidal particles building up a mesoscopically structured crystal which usually diffract X-rays like a single crystal.[34]

We already made the acquaintance with the principle of amorphous intermediates in biogenic mineralization processes. In principle, the occurrence of a transient amorphous phase is fits in the Ostwald's rule of stages.[35] This rule predicts that "[...] in the course of transformation of an unstable (or metastable) state into a stable one the system does not directly go to the most stable conformation (corresponding to the modification with the lowest free energy) but prefers to reach intermediate stages (corresponding to other metastable modifications) having the closest free energy to the initial state".[36] The Ostwald-Volmer rule varies this a little bit by stating that the least stable and least dense modification is chosen. Observation of amorphous precursors is still difficult since they are highly instable and transform quickly. But by increasing knowledge of their existence, their observation is today quite numerous: they were efficiently traced in different studies, *e. g.* by transmission electron microscopy, fast drying, small- and wide-angle X-ray scattering, X-ray absorption spectroscopy or X-ray microscopy.[37–46]

Amorphous Precursors

A very special case of amorphous intermediates are liquid intermediates. Being very instable, they tend strongly to re-dissolution, solidification or crystallization. The existence of liquid mineral precursors seems to be restricted to calcium carbonate based systems since no other mineral system was reported to feature such an intermediate so far. This may be due to the challenging experimental condition which have to be applied: even approaches like fast-mixing are full of uncertainty.[47] Wegner *et al.* were the first who proposed a liquid calcium carbonate precursor and discussed further the probable spinodal liquid/liquid phase separation of a supersaturated calcium carbonate solution.[44,45,48] In 2004, they suggested a lower critical solution temperature of 10° C which should characterize the proposed spinodal phase diagram (*cf.* Fig. A.3 in the appendix).

Gower reported earlier that a liquid precursor can be induced by little amounts of small anionic polymers and was thus termed 'polymer-induced liquid precursor' (PILP). The employed acidic polymers were suggested to sequester the cations, increase the water-content of the forming mineral phase and delay considerably the crystallization.[49,50] This liquid precursor phase has a great potential to be employed as an extraordinary building material.

1. Introduction: Nonclassical Crystallization in vivo et in vitro

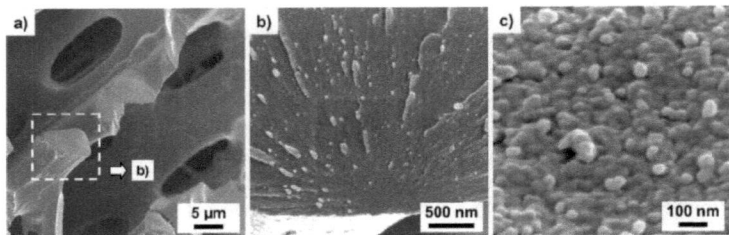

Figure 1.3 Scanning electron micrographs of the nanosized building blocks of an *Echinoderm* skeleton as presented by Oaki et al.:[56] **(a)** The fractured surface shows no cleavage planes but a conchoidal fracture surface. **(b)** The building blocks are radially orientated and **(c)** are of nanosize.

The liquid calcium carbonate droplets of roughly 100 nm may be molded in any shape and keep their shape under transformation to a crystalline material.[51] The PILP can be used for coating by sedimentation, coalescence and finally crystallization.[50] The liquid state of this kind of intermediate opens up an excellent opportunity for demanding molding of non-equilibrium morphologies, even if the emerging materials are polycrystalline.

Meso-crystallization The very special crystallization pathway of mesocrystallization cannot be described within the picture of the classical Ostwald ripening, *i.e.* the growth of larger crystals at the expense of smaller ones.[52] Most of the concept of mesocrystallinity was developed by Cölfen and coworkers.[53,54] They define mesocrystals as "colloidal crystals that are build up from individual nanocrystals and are aligned in a common crystallographic register" and "usually scatters X-rays like a single crystal".[34] As Cölfen himself stated, this definition may be too strict but deals as a good conceptional outline. Since they are build up from smallest building blocks, mesocrystals and their one- and two-dimensional relatives can adopt any shape. Typically, their formation is based on an interplay between the nanocrystals and a stabilizer which prevents dissolution, further crystallization or a 'coalescence' of the mesocrystal to a single crystal.[55]

Once again, Nature provides an excellent example of a mesocrystallization in the form the sea urchin spicule (*phylum Echinodermata*). By wide-angle X-ray analysis, the spicule turn out to be a single-crystalline calcite and these findings were seconded polarization microscopy. If it is fractured, no typical cleavage planes come up but a conchoidal fracture surface, which is typical for glassy or amorphous bodies. Weiner *et al.* proposed that this fracture behavior is due to occluded protein.[57] Recent studies of Oaki *et al.* substantiated that the transformation of ACC to one single-crystalline spicule can be understood in fact as a mesocrystallization with occluded proteins acting as stabilizing acidic polymer (*cf.* Fig. 1.3).[56] All characteristics of sea urchin spicules are consistent with the definition of a mesocrystal: building blocks are nanosized, radially aligned and the complete composite behaves like a single crystal. This is not the sole example of biological mesocrystallinity since mutually aligned building blocks were also found in nacre.[58–61] Thus, mesocrystallinity is not only beneficial for advanced morphogenesis but may open up another opportunity for engineering tailored composite hybrid materials.

References

[1] F. W. Tegethoff (ed.), *Calcium carbonate: from the Cretaceous period into the 21. century*, Birkhäuser, Basel, Boston, Berlin **2001**, ISBN 978-3764364250.

[2] *The Economics of Ground Calcium Carbonate*, Roskill Information Services, 3rd ed. **2007**, ISBN 978-0-86214-549-1.

[3] *The Economics of Precipitated Calcium Carbonate*, Roskill Information Services, 7th ed. **2008**, ISBN 978-0-86214-545-3.

[4] E. Bäuerlein, *Biomineralization*, Wiley–VCH, Weinheim **2007**, ISBN 978-3-527-31641-0.

[5] Aristotle, *Metaphysics*, Penguin **1998**, ISBN 0-14-044619-2, translated by H. Lawson–Tancred.

[6] S. Mann, *Biomineralization*, Oxford University Press, Oxford **2001**, ISBN 0-19-850882-4.

[7] H. A. Löwenstam and S. Weiner, *On Biomineralization*, Oxford University Press, New York **1989**.

[8] E. Haeckel, *Kristallseelen – Studien über das anorganische Leben*, Vdm Verlag Dr. Müller **1917 / 2006**, ISBN 978-3-86550-927-4.

[9] J. D. H. Donnay and D. Harker, *Amer Miner* **1937**, *22* (5), 446.

[10] P. Hartman and W. G. Perdok, *Acta Cryst* **1955**, *8*, 46.

[11] P. Hartman and P. Benema, *Acta Cryst* **1980**, *49*, 145.

[12] G. Hofmann (ed.), *Kristallisation in der industriellen Praxis*, Wiley–VCH **2004**, ISBN 3-527-30995-0.

[13] S. B. Mukkamala and A. K. Powell, *Chem Commun* **2004**, 918.

[14] V. Goldschmidt, *Atlas der Krystallformen*, Carl Winters Universitätsbuchhandlung, Heidelberg **1913**.

[15] E. Haeckel, *Kunstformen der Natur*, Prestel Verlag, München, New York **1904 / 1998**, ISBN 3-7913-1978-7.

[16] J. R. Young, M. Geisen, L. Cros, A. Kleijne, C. Sprengel, I. Probert and J. Østergaard, *J Nannoplankton R* **2003**, (1), special issue. Also available online: *http://www.nhm.ac.uk/hosted_sites/ina/CODENET/GuideImages/*.

References

[17] L. Addadi, S. Raz and S. Weiner, *Adv Mater* **2003**, *15*, 959.

[18] E. Beniash, J. Aizenberg, L. Addadi and S. Weiner, *Proc R Soc London Ser B* **1997**, *264*, 461.

[19] I. M. Weiss, N. Tuross, L. Addadi and S. Weiner, *J Exp Zool* **2002**, *293*, 478.

[20] S. Raz, P. C. Hamilton, F. H. Wilt, S. Weiner and L. Addadi, *Adv Funct Mater* **2003**, *13*, 480.

[21] Y. Politi, T. Arad, E. Klein, S. Weiner and L. Addadi, *Science* **2004**, *306*, 1161.

[22] Y. Ma, S. Weiner and L. Addadi, *Adv Funct Mater* **2007**, *17*, 2693.

[23] J. L. Arias, A. Neira-Carrillo, J. I. Arias, C. Escobar, M. Bodero, M. David and M. S. Fernandez, *J Mater Chem* **2004**, *14*, 2154.

[24] J. Aizenberg, G. Lambert, L. Addadi and S. Weiner, *Adv Mater* **1996**, *8* (3), 222.

[25] R. S. K. Lam, J. M. Charnock, A. Lennie and F. C. Meldrum, *CrystEngComm* **2007**, *9*, 1226.

[26] Y. Levi-Kalisman, S. Raz, S. Weiner and L. Addadi, *Adv Funct Mater* **2002**, *12* (1), 43.

[27] K. Wada, *Bull Natl Pearl Res Lab* **1961**, *7*, 703.

[28] A. M. Belcher, X. H. Wu, R. J. Christensen, P. K. Hansma, G. D. Stucky and D. E. Morse, *Nature* **1996**, *381*, 56.

[29] J. B. Thompson, G. T. Paloczi, J. H. Kindt, M. Michenfelder, B. L. Smith, G. Stucky, D. E. Morse and P. K. Hansma, *Biophys J* **2000**, *79*, 3307.

[30] S. Mann (ed.), *Biomimetic Materials Chemistry*, Wiley–VCH, New York **1997**, ISBN 978-0-471-18597-0.

[31] S. Mann, *Nature* **1993**, *365*, 499.

[32] F. Meldrum, *Int Mater Rev* **2003**, *48*, 187.

[33] J. Aizenberg, D. A. Muller, J. L. Grazul and D. R. Hamann, *Science* **2003**, *299* (5610), 1205.

[34] H. Cölfen and M. Antonietti, *Mesocrystals and Nonclassical Crystallization*, John Wiley & Sons Ltd, The Atrium, Southern Gate, Chichester **2008**, ISBN 987-0-470-02981-7.

[35] W. Ostwald, *Z Phys Chem* **1897**, *22*, 289.

[36] J. Schmelzer, J. Moeller and I. Gutzow, *Z Phys Chem* **1998**, *204* (1/2), 171.

[37] H. Cölfen and L. M. Qi, *Chem Eur J* **2001**, *7* (1), 106.

[38] J. Rieger, T. Frechen, G. C. W. Heckmann, C. Schmidt and J. Thieme, *Faraday Discussions 136 on Nucleation and Crystal Growth*, Roy. Soc. Chem., London, 265–278.

[39] J. Rieger, J. Thieme and C. Schmidt, *Langmuir* **2000**, *16* (22), 8300.

[40] J. Rieger, E. Hadicke, I. U. Rau and D. Boeckh, *Tens Surf Det* **1997**, *34* (6), 430.

[41] D. Pontoni, J. Bolze, N. Dingenouts, T. Narayanan and M. Ballauff, *J Phys Chem B* **2003**, *107* (22), 5123.

[42] J. Bolze, B. Peng, N. Dingenouts, P. Panine, T. Narayanan and M. Ballauff, *Langmuir* **2002**, *18* (22), 8364.

[43] M. Faatz, W. Cheng, G. Wegner, G. Fytas, R. Penciu and E. Economou, *Langmuir* **2005**, *21* (15), 6666.

[44] M. Faatz, F. Gröhn and G. Wegner, *Mater Sci Eng C* **2005**, *25* (2), 153.

[45] M. Faatz, F. Gröhn and G. Wegner, *Adv Mater* **2004**, *16* (12), 996.

[46] M. Balz, H. Therese, J. Li, J. S. Gutmann, M. Kappl, L. Nasdala, W. Hofmeister, H.-J. Butt and W. Tremel, *Adv Funct Mater* **2005**, *15* (4), 683.

[47] H. Haberkorn, D. Franke, T. Frechen, W. Goesele and J. Rieger, *J Coll Interface Sci* **2003**, *259* (1), 112.

[48] M. Faatz, *Kontrollierte Fällung von amorphem Calciumcarbonat durch homogene Carbonatfreisetzung*, Ph.D. thesis, Johannes Gutenberg – Universität, Mainz **2005**.

[49] L. B. Gower and D. A. Tirrell, *J Cryst Growth* **1998**, *191*, 153.

[50] L. B. Gower and D. J. Odom, *J Cryst Growth* **2000**, *210* (4), 719.

[51] R. W. Gauldie and D. G. A. Nelson, *Compar Biochem Physiol A – Physiol* **1988**, *90*, 501.

[52] W. Ostwald, *Lehrbuch der Allgemeinen Chemie*, Leipzig, **1896**, Germany.

[53] S. Wohlrab, N. Pinna, M. Antonietti and H. Cölfen, *Chem Eur J* **2005**, *11*, 2903.

[54] M. Niederberger and H. Cölfen, *Phys Chem Chem Phys* **2006**, *8*, 3271.

[55] S. Mann, *Angew Chem Int Ed* **1999**, *39* (19), 3392.

[56] Y. Oaki and H. Imai, *Small* **2005**, *2* (1), 66.

[57] S. Weiner, L. Addadi and H. D. Wagner, *Mater Sci Eng C* **2000**, *11*, 1.

[58] Y. Oaki and H. Imai, *Angew Chem Int Ed* **2005**, *44*, 6571.

[59] M. Rousseau, E. Lopez, P. Stempfle, M. Brendle, L. Franke, A. Guette, R. Naslain and X. Bourrat, *Biomaterials* **2005**, *26*, 6254.

[60] Y. Dauphin, *Palaontolog* **2001**, *75*, 113.

[61] X. D. Li, W. C. Chang, Y. J. Chao, R. Z. Wang and M. Chang, *Nano Lett* **2004**, *4*, 613.

References

2 Research Objectives

This contribution will explore the occurrence of nonclassical crystallization processes focussing on the formation of calcium carbonate and related bivalent carbonates.

The occurrence of liquid calcium carbonate intermediates prior to the formation of crystalline material is still questionable. Several publications deal with the proposition or report of a liquid intermediate, but mostly the reports suffer from drawbacks like potential experimental artifacts or simply the lack of experimental evidence. In case of the 'polymer-induced liquid-precursor' model proposed by L. A. Gower, the existence of a liquid calcium carbonate precursor was proven clearly and repeatedly. Nevertheless the related detailed mechanism of formation still remains nebulous. The main objective of this contribution is to introduce an experimental setup which enables one to investigate in-depth the very early stages of homogenous calcium carbonate formation. Based on this setup, the existence of *liquid intermediates* during calcium carbonate formation is to be evidenced. As a following and generalizing step, carbonate minerals of other bivalent metals will be assayed whether they feature a liquid intermediate as well. The second main objective is to elucidate the *stabilization and formation mechanism of the liquid precursor*, be it polymer-induced or additive-free. The question is to be answered whether polymer-induced and additiv-free intermediates originate from the same chemical phenomenon.

(i) Liquid Intermediates...

... and their Mechanisms of Stabilization and Formation?

The topic of nonclassical crystallization comprises more than the occurrence of amorphous or liquid intermediates. Two further studies will complete the picture of nonclassical crystallization of calcium carbonate. The molding of calcium carbonate can be efficently achieved by employing principles of *mesocrystallinity*. This was evidenced by Cölfen *et al.* in numerous and elegant studies. In these studies, the mineral is the majority part; it is the material the mescrystal is build of. But the intracrystalline minority constituent, *e. g.* occluded protein in the nacrous layer of *Haliotis laevigata*, experiences as well molding. Strictly speaking it is exposed to compression-molding. This aspect of mesocrystallization takes today second place to the more obvious molding of the majority mineral component. Is it possible to employ the concept of compression-molding during mesocrystal formation to mold a mineralic component?

(ii) Compression-Molding in Mesocrystals?

2. Research Objectives

(iii) Symmetry-Breaking Crystallization? A crystallization violating not the classical theories of crystal birth and growth but the classical rules of symmetry surly has to be ranked among nonclassical crystallization. A *symmetry-breaking* crystallization of calcium carbonate will serve as the concluding and most uncommon example of nonclassical crystallizations while addressing the issue of symmetry-breaking phase selection.

3 Nonclassical Liquid Intermediates During the Homogenous Formation of Calcium Carbonate

Abstract Homogenous formation of calcium carbonate was followed contact-free and *in situ* in acoustically levitated droplets. The employed levitation technique allows *in situ* monitoring of the crystallization while avoiding any foreign phase boundaries that may influence the precipitation process favoring heterogenous nucleation. Precipitation of $CaCO_3$ was conducted diffusion-controlled at neutral pH and starts in the initial step with the homogenous formation of a nanosized liquid amorphous calcium carbonate emulsion. This transient phase is apparently stabilized electrostatically and undergoes in a subsequent step a solution-assisted transformation to calcite. Cryogenic scanning electron microscopy studies evidenced that the precipitation is not induced at the solution/air interface. These findings demonstrate that a liquid/liquid phase separation occurs at the outset of the precipitation under diffusion-controlled conditions with a slow increase of the supersaturation at neutral pH, which is typical for biomineral formation.

Associated publications	*J Am Chem Soc* **2008**, *130* (37), 12342–7. *Nanoscale* **2011**, *3*, 1158–1165.
Highlighted in	*Nachr Chemie* **2008**, *10*, 90.

Introduction

The formation of calcium carbonate has been studied for more than a century with more than 3000 papers over the past 10 years. Whereas the undesired precipitation of calcium carbonate is a persistent, expensive, and widespread problem,[1–4] the directed formation of calcium carbonate with controlled particle size, shape, and crystallographic phase is crucial for various commercial applications. Crystallization of calcium carbonate in natural environments may be summarized under the label biomineralization (*cf.* Chp. 1), where precipitation occurs in the presence of large organic molecules resulting in the formation of hierarchically ordered inorganic-organic hybrid structures.[5-7] Recently, much attention has been devoted to amorphous calcium carbonate (ACC) as a singular material because evidence is increasing that this phase plays a crucial role in biomineralization. ACC is the most unstable form of calcium carbonate, and under ambient conditions it transforms quickly into more stable crystalline forms, such as vaterite and calcite.[8,9] Many mineralization processes are now believed to envolve the transformation of a transient amorphous precursor,[10] which has been shown to act a reactive in intermediate in generating complex functional materials.[11,12]

Various analytical methods such as fast drying,[13] cryogenic transmission electron microscopy (cryo-TEM),[14,15] small- and wide-angle X-ray scattering (SAXS, WAXS)[16-18] and X-ray microscopy[19] have been utilized to observe the initial formation steps. Rieger et al. studied the formation of calcium carbonate at high supersaturation ($c \approx 0.01$,mol l^{-1} during precipitation) after rapid mixing of the reactants $CaCl_2$ and Na_2CO_3. Cryo-TEM studies revealed the formation of emulsion-like structures preceding the precursor stage and triggered speculations about a spinodal phase separation between a denser and a less dense phase.[15] Faatz et al. reported the formation of calcium carbonate from a reaction of calcium chloride with carbon dioxide, which was homogenously released to the solution by alkaline hydrolysis of alkyl carbonate. This homogenous formation of CO_2 in the reaction medium prevents the formation of a gas/liquid interface and the formation of amorphous calcium carbonate is postulated to proceed by a liquid/liquid binodal phase separation mechanism,[20,21] but no analytical support for the formation of the proposed emulsion-like early stages could be provided. Navrotsky explains the formation of monodisperse nanoparticles within the classical La Mer nucleation and growth model.[22,23]

The main disadvantage of the most experimental approaches to the precipitation of calcium carbonate is the diverse and poorly defined precipitation conditions. Firstly, by way of the heterogenous dissociation equilibrium of carbonic acid the pH plays an important role. Furthermore, in most studies precursor solutions were rapidly mixed under turbulent conditions in order to achieve a sufficiently large supersaturation. There are severe disadvantages of the rapid mixing approach. The system starts reacting at the interface of two intermixing liquid educts, but a state of homogenous supersaturation is not reached.[24] As a result, artifacts can occur, e. g. an instantaneous reaction at the interfaces of the two reactants that meet in the mixing device.[25] From this preliminary state, primary particles form with sizes in the nanometer range.

Introduction

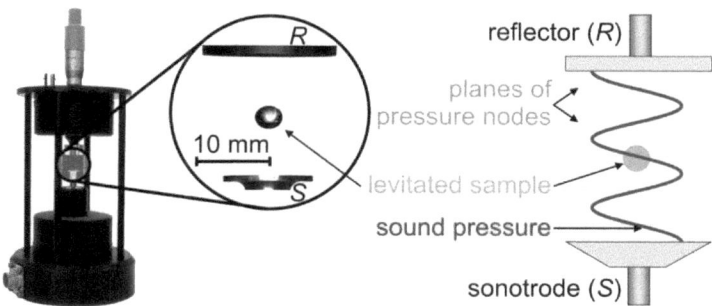

Figure 3.1 Photograph (left) and mode of operation (right) of the acoustic levitator. A droplet of water is levitated in the central node of the standing acoustic wave (inset) between the sonotrode S and a concentrically adjusted reflector R to demonstrate the levitation of a liquid sample.

Nucleation is strongly affected by foreign bodies (*e. g.* macromolecules, spectator ions, liquid/liquid- or solid/liquid-interfaces, like vessel walls or due to mixing processes).[26] The potential barrier of homogenous nucleation, which must be overcome for the formation of a new phase, is a function of the interfacial energy between the crystal and its mother phase. Refereing to this point, foreign materials and their phase boundaries will have an impact not only on the nucleation and crystal growth rates, but also on the particle size distribution of the product. They can induce a shift away from a homogenous nucleation mechanism by energetically favoring a heterogenous path. Consequently, it is very difficult to study homogenous nucleation respective phase separation without intervening heterogenous bodies *e. g.* walls of a nozzle or a mixing chamber. It is still a matter of controversy, whether genuine homogenous nucleation can be observed at all, as it is almost impossible to remove any foreign body or phase boundary from a system.

Actually, acoustic levitation permits a contact-free crystallization with a minimum of perturbations and requirements for the levitated liquid samples,[27] and the occuring processes can be monitored *in situ* by synchrotron X-ray scattering techniques. The solution/air interface remains as the only phase boundary. No other contact or foreign materials are present. In a levitated droplet, the analyte mass is constant during evaporation and no losses occur. Working at an oscillating frequency of 58 kHz, a piezoelectric vibrator acts as an ultrasonic radiator (see Fig. 3.1).[28,29] A standing acoustic wave is generated between this sonotrode and a concentrically adjusted reflector at a distance of some multiple of half the wavelength. As a result of axial radiation pressure and radial Bernoulli stress, liquid and solid samples can be held in a levitated and contact-free position by placing it in sound pressure nodes (for further details, see Sec. 8 in the appendix). Typically, levitated samples have a volume of 5 nL – 5 µL, which corresponds to a sample diameter of 0.2 – 2 mm. No other constraints on the sample, such as magnetic or dielectric properties, apply for acoustic levitation.

Apart from the ultrasonic levitation method applied here, only few other methods are applicable for an *in situ* monitoring of crystallization and phase transition processes under contact-free conditions. But either special requirements have to be met or the methods are not contact-free in a strict sense. Electromagnetic levitation require electrical conductivity of the samples,[30,31] whereas methods based on electrostatic levitation operate under vacuum.[32] Free-jet methods require a fast mixing of educts before generating a steady-state jet of the intermixed liquids through a nozzle; the intermediates are captured by cryo-TEM.[25] This combination of methods implicates various possible artifacts: *(i)* The mixing may be incomplete due to turbulent conditions as discussed above.[25] *(ii)* The precipitation is not contact-free in a strict sense and suffers from wall contacts of the mixing chamber. *(iii)* Rieger et al. stated some uncertainties in sample preparation for cryo-TEM, such as differences in (local) cooling rates.[25]

Results   A droplet of an aqueous saturated solution of calcium bicarbonate with a volume of roughly 4 µL was injected in an ultrasonic levitator. The calcium bicarbonate concentration increases due to the evaporation of water, and calcium carbonate starts precipitating because carbon dioxide is released. Starting from sub-critical concentrations regarding precipitation and supersaturation, turbulent mixing conditions of the educts are strictly avoided, and the influence of foreign phase boundaries except for the air/solution-interface (*e. g.* vessel walls during preparation or injection into the levitator) can be ruled out. The formation and growth of calcium carbonate in levitated droplets was followed time-resolved from undersaturated to supersaturated concentrations in a single experiment by *in situ* WAXS experiments performed at a synchrotron micro-spot beamline. In addition, different stages of the crystallization were characterized by transmission electron microscopy as well as by cryogenic and standard scanning electron microscopy (cryo-SEM and SEM).

The mineralization was monitored by time-resolved WAXS. The respective diffraction patterns are presented in Fig. 3.2. The early patterns show only the diffuse scattering of water which vanishes gradually as the water evaporates. The first detectable reflection belongs to the {104} set of the calcite lattice planes; its intensity increases throughout the experiment. The other calcite reflections (102), (110), (113), and (202) are detectable after 22 min, whereas the weak (006) reflection was observed after 34 min. The absence of

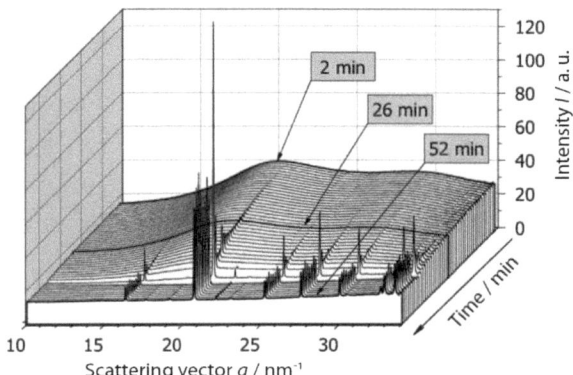

Figure 3.2 Scattering curves recorded during evaporation of a levitated saturated solution of $Ca(HCO_3)_2$ which lasted 1 h. Each diffraction pattern was recorded with an exposure time of 40 s.

a preferred orientation of the calcite crystals can be deduced from the spot-free and continuous Debye-Scherrer rings in the 2D frames (see Fig. 3.3a). A minimum of the mean particle diameter can be estimated to approximately 110 nm from the Scherrer equation based on the (012) reflection at lowest q (for more details, *cf.* Experimental Part, p. 28).[33] These values correspond with particle diameters obtained from TEM and cryo-SEM (*vide infra*).* *In situ* WAXS experiments do not indicate the presence of crystalline phases other than calcite in significant amounts; traces of vaterite were observed only in the final stages of the droplet evaporation ($< 5\%$ based on the results of Rietveld refinements).[35]

As mentioned above, these early amorphous reaction stages were characterized by transmission electron microscopy (TEM). The primary product consists of nanospheres with an emulsion-like appearance (Fig. 3.4a and 3.4b). The low contrast variation within the particles indicates their liquid-like character. Solid spherical particles would show a distinct increase in contrast from the surface to the center of the particles. Particle diameters range from 100 to 300 nm, and their amorphous state

---

* For higher scattering vectors the integral width of the reflections increases monotonically resulting in an apparent smaller calculated size of the crystallites (see Experimental Part for further details). The advantage of a determination of particle diameters employing the Scherrer method lies in the possibility to obtain average values for large quantities, whereas particle sizes determined by TEM or SEM are (ideally) average values for a small number (≈ 50) of particles. See Borchert *et al.*[34]

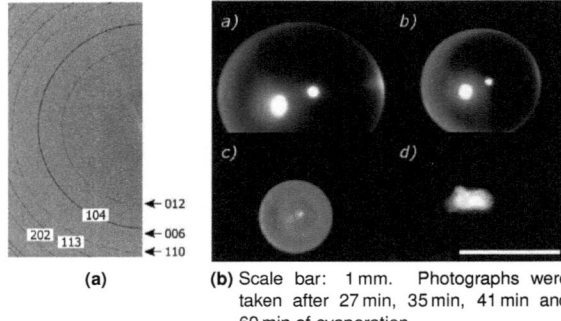

Figure 3.3 (a) Representative 2D detector frame measured in a late state of experiment. (b) Microscopy images of the evaporation of a levitated droplet of a Ca(HCO$_3$)$_2$ solution. The whole droplet appears increasingly opalescent due to an aggregation of nanoparticles. The very bright spot close to the center of the droplet is due to reflected light. Scale bar: 1 mm. Photographs were taken after 27 min, 35 min, 41 min and 60 min of evaporation.

was confirmed by electron diffraction (ED, see inset in Fig. 3.4b). Remarkably, these amorphous particles are stable without stabilizing surfactants for several hours up to a few days. After long storage or upon radiation damage, crystallization commences. If artificial nuclei (*e. g.* gold nanoparticles) are present, particles with a deviant, non-liquid appearance were formed (Fig. 3.5 and 3.6).

During the early stages of the crystallization, all Bragg reflections are split. As an example the (104) reflection profile at $q = 20.68\,\text{nm}^{-1}$ and $20.79\,\text{nm}^{-1}$ is shown in Fig. 3.7. This doubling of reflections is caused by diffraction from two separate areas of the levitated droplet due to differences in the distance between detector and sample. According to the geometrical relations given in the Experimental Part (*cf.* p. 28), one calculates a distance of 1.52 mm. This finding shows that the phase transformation process from amorphous calcium carbonate to crystalline calcite starts at the droplet surface and thus clearly differs from the crystallization within the droplets.[27]

In order to determine, whether the amorphous particles initially form heterogenously at the air/water-surface or homogenously in solution, a droplet was vitrificated in liquid ethane, fractured, and the frozen droplet fragments were dried slowly by ice sublimation. Cryo-SEM revealed that these first particles form homogenously within the droplet volume (Fig. 3.8). The particle diameter is in perfect agreement to those determined by TEM. When a droplet was levitated for about 400 s, a 20 µm thick layer was formed at the droplet surface. The particle density in this surface layer is much higher than particle number density within the volume of the droplet. As the droplet

Results

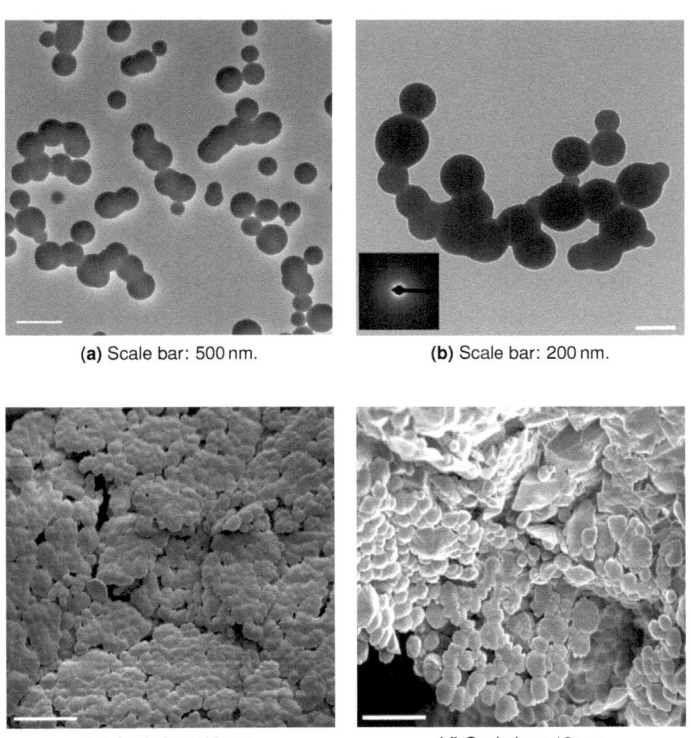

(a) Scale bar: 500 nm.   (b) Scale bar: 200 nm.

(c) Scale bar: 20 µm.   (d) Scale bar: 10 µm.

Figure 3.4 **(a, b)** TEM micrographs of calcium carbonate particles obtained after 400 s. The low contrast variation within the particles indicates their liquid character. **(c, d)** After complete evaporation, SEM revealed that spherical solid particles are present along with rhombohedral calcite crystals; one single calcite crystal is marked in (d). Based on the morphology of the calcite particles, the transition to the crystalline phase is assumed to take place through recrystallization.

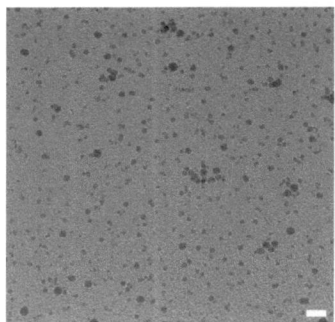
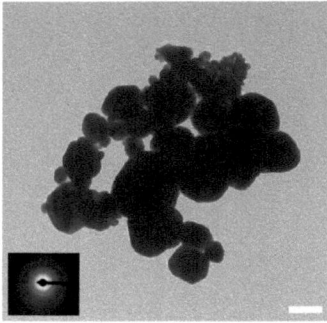

(a) 300 s; scale bar: 100 nm.   (b) 400 s; scale bar: 100 nm.

Figure 3.5 Transmission electron micrographs of crystalline calcium carbonate particles obtained in presence of gold nanoparticles after (a) 300 s and (b) 400 s.

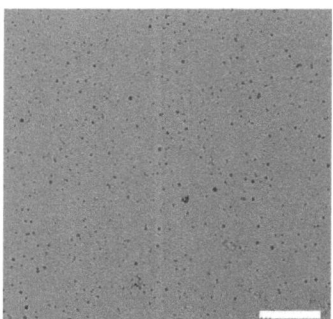

(a) 300 s; scale bar: 500 nm.   (b) 400 s; scale bar: 100 nm.

Figure 3.6 Transmission electron micrographs of crystalline calcium carbonate particles obtained in presence of calcium carbonate nanoparticles after (a) 300 s and (b) 400 s.

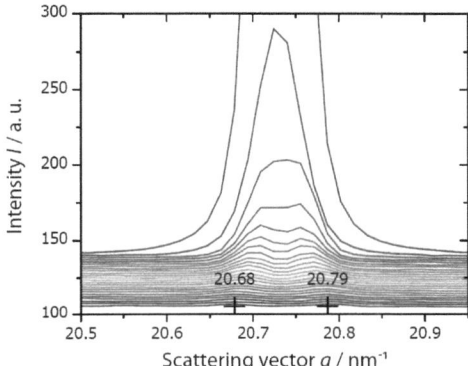

Figure 3.7 Evolution of the profile of the (104) reflection with increasing sample concentration (from the lower blue to the upper red curve). Each curve is shifted by 20 a. u. in direction of I for illustration, and were normalized to $I = 100$ at $q = 21.5\,\text{nm}^{-1}$. The reflex at higher scattering angle is more intense than the one at lower angle. The latter one derives from crystallites located at the front side of the droplet and the peak at higher angle from its backside. As a result of additional absorption and scattering losses this peak is less intense.

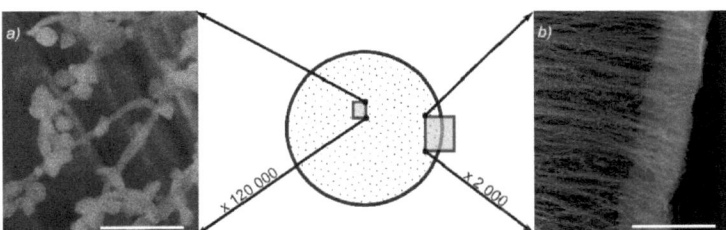

Figure 3.8 Cryo-SEM studies on a droplet levitated for 400 s, which was vitrificated in liquid ethane, then cryo-fractured and lyophilized. Particles form in the whole droplet volume. Their diameter is in good agreement with diameters derived from TEM studies. **(a)** Particles found in the inner part of the droplet are likely to be formed homogenously. **(b)** A 20 µm thick layer near the droplet surface develops because particles accumulate close to the surface while the droplet shrinks by evaporation. The fibrous structures in both images are artifacts due to ice devitrification. Scale bars: (a) 500 nm (b) 40 µm.

shrinks due to solvent evaporation, the already particles accumulate close to the droplet surface. The sharp inner boundary without any gradient of the particle density clearly indicates that the particles are only collected at and not formed within the surface layer. This effect may be viewed as a three dimensional variant of the so-called coffee-stain effect.[36,37]

Scale factors s, which can be extracted from a Rietveld analysis of the X-ray powder data series,[33] represent the transformed volume fraction $f_V(t)$ of the calcite phase when all calcite scale factors are rescaled to run between zero and one. A non-linear least-square fit of this evolution to the Avrami form (Eq. 3.1 and *cf.* Sec. 8 in the appendix) yields a value for the Avrami exponent n that mainly characterizes the kinetics of the phase formation.[38–41]

$$f_V(t) = 1 - e^{-k_A(t-t_0)^n} \qquad \text{(Eq. 3.1)}$$

For polymorphic transitions, $n \approx 3$ indicates a mechanism where nucleation occurs only at the start of a transformation, a value $n \approx 4$ indicates that the phase transformation continues to form new growing nuclei in untransformed material. A non-linear least-squares fit to the Avrami form of the present calcite phase yields a value for the exponent $n = 8.3$, with $t_0 = 26$ min as the time, when crystallization is first seen as a Bragg reflection (*cf.* Fig. 3.9). This large value of $n = 8.3$ seems to indicate that the transformation rate to calcite is high, and it implies a secondary nucleation on amorphous calcium carbonate particles independent of the calcite crystallites that already exist. Thus, the observed precipitation of calcite occurs rapidly once a critical level has been reached after approximately 26 min. The analysis of the dry residue by scanning electron microscopy (SEM) showed spherical particles with diameters of about 5 μm, together with several crystals exhibiting the rhombohedral morphology of calcite crystals (Fig. 3.4c and d). The observed spherical particles are assumed to be solidified dry amorphous calcium carbonate which did not transform into crystalline material.

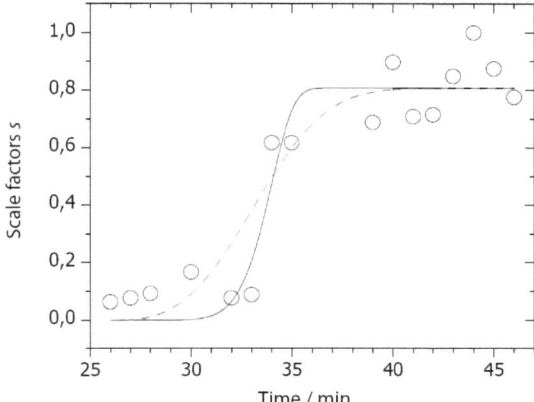

Figure 3.9 Time evolution of normalized scale factors s(t) from selected Rietveld refinements (circles) non-linear least-square fitted to the Avrami equation (solid line) yielding $n = 8.3$, $t_0 = 26$ min and $k_A = 0.123$ min^{-1}. For comparison, a simulated curve with the parameter $n = 3$ is given (dashed line).

The supersaturation of calcium carbonat eincreases due to the loss of water by evaporation and a concomitant loss of carbon dioxide from the droplet. Primary particles are amorphous and of homogenous origin. Cryogenic SEM analysis proved that the particle formation is not induced heterogenously at the air/water interface. The intermediate calcium carbonate phase possess an emulsion-like structure, and its occurrence is compatible with a homogenous liquid/liquid phase separation process. Upon aging, the particles continuously lose water and solidify, and amorphous calcium carbonate particles accumulate close to the surface due to the shrinking of the droplet. This thin layer of particles gives rise to the opalescent appearance of the droplet.

The fact that no other calcium carbonate phase appears seems to contradict Ostwald's rule of stages,[42] which states that the next phase to appear will be always the nearest by state in energy (*cf.* p. 5). Our findings, however, can be rationalized by a templated nucleation of calcite on the surface of the amorphous precursor.[43] The accumulation of amorphous particles in the thin surface layer near the droplet surface provides a large surface area for secondary heterogenous nucleation. The incipient calcite formation leads to the observed peak splitting in the early stages of crystallization (*vide supra* and Fig. 3.7).

Calcite formation increases rapidly after approximately 26 min. The high nucleation rate deduced from the Avrami plot goes along with a heterogenous nucleation of calcite, the crystallographically detectable phase, as well. A dissolution-assisted route seems feasible due the well-developed shape of the calcite crystals. The amorphous calcium carbonate phase does not transform completely into calcite. For the formation of calcium carbonate, a binodal liquid/liquid phase separation process was proposed by Faatz et al., but no analytical evidence was given.[20] Rieger et al. reported the formation of emulsion-like particles from a rapid mixing at high supersaturation. In this work it was demonstrated for the first time by a combination (X-ray, TEM, cryo-SEM) of methods that a liquid/liquid phase separation of calcium carbonate does not only occur for large supersaturations in case of a fast mixing processes at high pH, but also for a diffusion-controlled precipitation involving a slow increase of supersaturation at neutral pH. These conditions are close to the formation conditions of biominerals. The homogenous approach, based on the Kitano-Method without mixing of educts,[44] does not suffer from potential artifacts due to incomplete intermixing of the precursors.

The formation of calcium carbonate by a liquid precursor has been previously observed by Gower et al.[45] This 'polymer-induced liquid precursor' (PILP) process is believed to play an important role in the morphogenesis of biominerals and biomimetic materials. Little amounts of small poly-anionic polymers (such as poly-aspartate)[45] or proteins (e. g. ovalbumin, cf. Chp. 5) are thought to induce a liquid/liquid-phase separation during crystallization and small colloidal droplets of a metastable and amorphous liquid-phase mineral precursor are formed. The results presented here show that an amorphous liquid-phase mineral precursor is even formed in the absence of stabilizing polymers or additives, i. e. seems to be a characteristic feature of the homogenous formation of calcium carbonate itself, independent of the initial supersaturation. Polymers and proteins can aid in stabilizing this liquid state, but the liquid homogenous particles are remarkably stable at neutral pH as well.

Interestingly, the emulsion-like particles do not severely aggregate or coalesce in order to reduce their interfacial energy. This implies that the emulsions are electrostatically stabilized, as no other colloidal stabilization mechanisms apply. As described above, no spectator ions (e. g. Na^+ or Cl^- released during the reaction of Na_2CO_3 with $CaCl_2$) were present, whereas in almost all previous studies of the very early formation states of calcium carbonate such counter ions were present in high concentrations. Their presence is known to reduce the electrostatic repulsion forces and screen the surface potential that aggregation or coalescence may occur.[46] These conclusions are supported by experiments where the liquid calcium carbonate intermediate was prepared in the presence of 15 mmol l^{-1} of sodium chloride. In the presence of salt the calcium carbonate emulsion suffers from aggregation and coalescence (cf. Fig. 3.10).

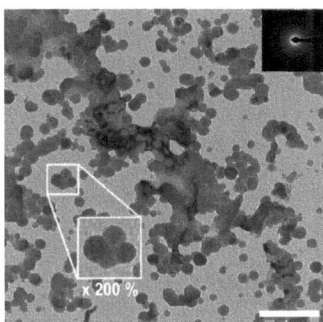

Figure 3.10 A liquid calcium carbonate intermediate prepared in presence of 15 mmol l^{-1} sodium chloride suffers from severe aggregation and coalescence, e.g. the 200-fold magnification inset shows a aggregate of four particles. This behavior points to an electrostatic stabilization of the emulsion. Scale bar: 500 nm.

The formation and stability of the liquid-like phase itself may be rationalized by the presence of several species involved in the process. The pH value exerts not only a strong influence on the supersaturation level by controlling the concentrations of carbonate, bicarbonate and non-dissociated carbonic acid. It also determines the concentration of the active calcium solution species that is involved in complexation and precipitation equilibria. Their hydroxyl groups can set up an extensive hydrogen-bonding network between the three carbonate species and water, which may interact with the hydrated or hydrogen carbonate-coordinated calcium ion. Therefore, at neutral pH, this network formation of several species may favor the formation of an amorphous solid (similar as in glass forming materials such as borates) over the formation an ionic calcium carbonate lattice. The presence of *(i)* various bonding partners, *(ii)* variable coordination geometries and coordination numbers of the carbonate groups and *(iii)* the associated distribution of local structures may favor the formation of the non-crystalline phase.

The ultrasonic trap was proven to serve as a powerful tool for a real-time analysis of nucleation, crystal growth, and phase separation processes by reducing disturbance and artifacts due to solid phase boundaries to a minimum. Thus, acoustic levitation provides a reliable sample environment for studies of homogenous precipitation reactions. Taking advantage of these benefits, the first *in situ* X-ray study of the contact-free homogenous precipitation of calcium carbonate in a levitated droplet has been conducted. Homogenous formation of calcium carbonate proceeds via an amorphous liquid-like state. This amorphous phase formed in the absence of any stabilizing polymers or additives at neutral pH, and the transient mineral emulsion is stabilized electrostatically. Its stability may be further related to the presence of various species such as carbonate, bicarbonate, and non-dissociated carbonic acid involved in the process. The formation of an amorphous liquid-phase mineral precursor seems to be a characteristic of the truly homogenous formation of calcium carbonate itself. In a second step, these primary particles template the crystallization of calcite. Finally, calcium carbonate may be regarded the first example where a liquid amorphous precursor could be identified for an inorganic mineral phase unambiguously without artifacts.[25]

Experimental Part

The crystallization was carried out homogenously according to the Kitano method.[44] By slow evaporation of water or a slow release of carbon dioxide from a saturated solution of calcium bicarbonate calcium carbonate is formed.

$$Ca(HCO_3)_2 \longrightarrow CaCO_3 + CO_2 + H_2O$$

A suspension of calcium carbonate (p. a., Sigma Aldrich) in ultra pure water (Millipore Synergy 185 with UV photo oxidation, 18.2 MΩ cm^{-1}) was extensively treated with carbon dioxide (Westfalen AG). The obtained saturated solution of calcium bicarbonate was filtered with a cascade of syringe filters, which consists of a 0.1 µm Millipore Millex VV followed by 20 nm Millipore Anotop in series. Afterwards, the filtered solution was again treated extensively with carbon dioxide to dissolve nuclei with diameters below 20 nm. In control experiments, crystallization nuclei were prepared by skipping this removal step in order to preserve calcium carbonate nuclei or by adding gold nanoparticles prepared according to the method of Gittins and Caruso.[47,48]

One droplet of an aqueous sample with a volume of approximately 4 µL was manually injected in the ultrasonic levitator (Tec 5, Oberursel, Germany). SEM investigations were performed with a Zeiss DSM 940 running at 10 kV. Samples for cryo-SEM were prepared by plunging the respective droplet in liquid ethane, and transferring it directly into a cryo-preparation chamber (FEI Quorum PolarPrep 2000), where it was cryo-fractured after removal of excessive ethane at −60° C and 10^{-6} mbar. Under these conditions, the vitreous ice transforms into a crystalline state (which is the origin of the fibrous artifacts mentioned above which are discussed in detail in Chp. 6). Afterwards the samples were transferred to the nitrogen-cooled cryo-stage of the cryo-SEM (FEI xT Nova 600 Nanolab), and the water was slowly removed by sublimation at −5° C and 10^{-6} mbar which was monitored *in situ* by SEM at 2 – 10 kV. TEM investigations were carried with a Phillips EM 420 running at 120 kV, equipped with an ORCA-ER Camera (1024 × 1024 pixel) and run with AMT Image Capture Engine v5.42.540a. The samples were

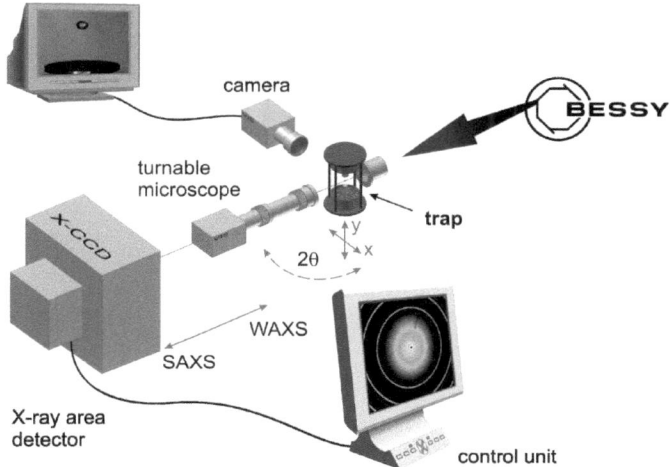

Figure 3.11 Experimental setup at the µSpot beamline integrates the acoustic levitator for SAXS and WAXS measurements. The levitated sample is monitored and remote controlled during the whole experiment. The pivoting microscope determines exactly the position of the 20 µm beam at the sample.

prepared by transferring the respective droplet on a lacey-coated TEM grid (Plano, Germany), followed by washing with water and air-drying.

Wide-angle X-ray scattering (WAXS) experiments were performed using the µSpot beamline at BESSY (Fig. 3.11), which is equipped with a double crystal monochromator, yielding a monochromatic X-ray beam of $\lambda = 1.00257$ Å. Further information concerning the general setup are given by Paris et al.[49] A pinhole system provides a beam of 20 µm cross section with a photon flux of about 10^9 per second and a ring current of 100 mA. The scattering pattern from corundum serves as external calibration standard. No mathematical 'desmearing' of the experimental scattering intensity function was needed due to the small beam diameter of the incident beam. In case of the levitated droplets, the scattering from the pure solvent was measured and used as an estimate of the background contribution. The obtained data was not corrected for background scattering, as it was not possible to correlate the shrinking volume of the sample solution with that of a water droplet for every state of evaporation considering the accompanying change of X-ray absorption. The scattering vector $q = 4\pi/\lambda \sin(\theta)$ is defined in terms of the angle 2θ between incident and scattered radiation of the wavelength λ. The data were processed and converted into diagrams of scattering intensities I versus q by employing algorithms of the computer program FIT2D.[50]

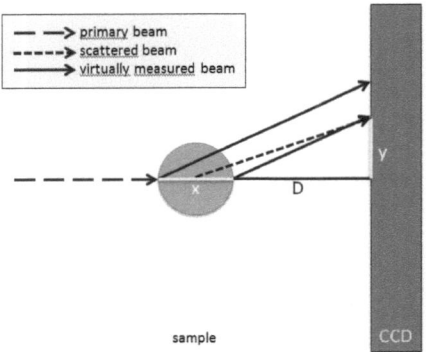

Figure 3.12 Geometric derivation of sample diameter.

Peak Splitting Based on the peak splitting, the sample diameter x can be extracted. In order to illustrate the relations in both triangles the shared opposite edge of the scattering angles θ_1 and θ_2 is marked (y) in Fig. 3.12. The distance between the detector and the sample is represented by D.

$$\tan \theta_1 = \frac{y}{D+x}$$
$$\tan \theta_2 = \frac{y}{D}$$
$$x = \frac{D \cdot \tan \theta_2}{\tan \theta_1} - D$$

(Eq. 3.2)

Scherrer Equation The Scherrer equation is given in the following equation, in which D_{hkl} is the crystallite size, k a constant close to one, λ the wavelength of the radiation used, ω is the peak width in 2θ scale and θ is the angle of the investigated Bragg reflection. It correlates on the wavelength independent q-scale through the derivation of the defined scattering vector q with respect to θ. The width of a reflection is defined by its integral width B_{int}.

$$D_{hkl} = k \frac{\lambda}{\omega \cos \theta}$$

(Eq. 3.3)

B_{int} makes the calculation of the crystallite size D_{hkl} independent on the reflection profile. Keeping in mind the negligible instrumental broadening

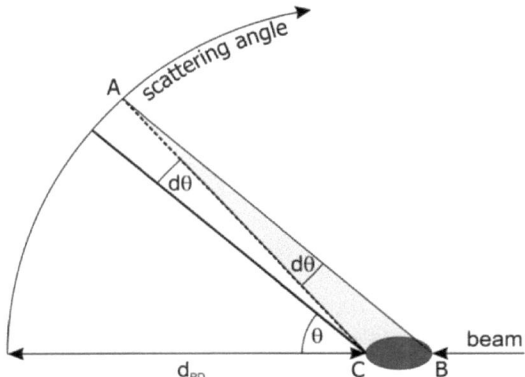

Figure 3.13 Geometric derivation of crystallite size.

of the reflection width, the reflection width of the diffraction patterns can be used to estimate the averaged grain size.

$$B_{int} = \frac{\int_0^\infty I(q)}{I(q_{max})} \qquad \text{(Eq. 3.4)}$$

The crystallite size was derived from diffraction pattern obtained from the final evaporated state, where no peak splitting was detected anymore. The derived integral width increases monotonically for higher scattering angles. Therefore, the mean value for the minimal size of crystallites decreases reciprocally. For synchrotron diffraction patterns the broadening due to beam divergence (< 1 mrad), energy resolution ($\Delta E/E = 10^{-4}$) and the diameter (20 µm) are negligible and will not be taken into account. Further influence of apparative broadening was taken into account by comparison to the external standard corundum. Furthermore, the observed sample diameter is small (r = 0.01 – 2 mm) in relation to its distance to the detector (d_{PD} = 267 mm), but large in relation to the beam diameter of 20 µm). Consequently, the analyzed scattering volume is limited by the beam diameter and the extent perpendicular to the beam, respectively. A further error in the direction of the beam is present. The maximum error appears due to diffraction on the sample surface for penetrating and leaving radiation after correction through absorption with maximum equal intensity. This error is at maximum for the smallest scattering angles and decreases for higher angles. The triangle ABC is defined by its two edges ($\overline{CA} = d_{PD}$ and $\overline{BC} = x$), its enclosed angle ($d\theta$), and the angle opposite to the longer edge (θ). Based on the law of sines and Fig. 3.13, the following straightforward relation can

be formulated.
$$\frac{x}{d_{PD}} = \frac{\sin d\theta}{\sin \theta}$$
$$d\theta = \arcsin \frac{x \sin \theta}{d_{PD}}$$

Here, x is the sample diameter and $d\theta$ is the broadening of the Bragg reflection. The broadening is negligible for the smallest scattering angles but increases for higher scattering angles and for larger samples as well. For the final state of the evaporation experiment, the broadening can be estimated to $B_{int} \approx 0.05\,\text{nm}^{-1}$. This is in the same range as the value of the measured reflection width and therefore no correction was applied. However, it causes an apparent decrease of the crystallite size for higher scattering angles. Anisotropic crystals show fluctuating values for different crystallographic axes. The values changes monotonously for the present measurements and therefore point to isotropic crystal forms.

References

[1] K. Ramstad, T. Tydal, K. M. Askvik and P. Fotland, Predicting carbonate scale in oil producers from high temperature reservoirs **2004**.

[2] Y. P. Zhang, H. Shaw, R. Farquhar and R. Dawe, *J Petr Sci Eng* **2001**, *29* (2), 85.

[3] J. Rieger, E. Hadicke, I. U. Rau and D. Boeckh, *Tens Surf Det* **1997**, *34* (6), 430.

[4] M. Kind, *Chem Eng Proc* **1999**, *38* (4-6), 405.

[5] H. A. Löwenstam and S. Weiner, *On Biomineralization*, University Press, Oxford **1989**.

[6] E. Bäuerlein, *Biomineralization*, Wiley–VCH, Weinheim **2007**, ISBN 978-3-527-31641-0.

[7] S. Mann, *Biomineralization*, Oxford University Press, Oxford **2001**, ISBN 0-19-850882-4.

[8] L. Addadi, S. Raz and S. Weiner, *Adv Mater* **2003**, *15*, 959.

[9] F. Meldrum, *Int Mater Rev* **2003**, *48*, 187.

[10] E. Beniash, L. Addadi and S. Weiner, *J Struct Biol* **1999**, *125* (1), 50.

[11] N. Gehrke, N. Nassif, N. Pinna, M. Antonietti, H. S. Gupta and H. Cölfen, *Chem Mater* **2005**, *17*, 6514.

[12] X. G. Cheng and L. B. Gower, *Biotechnol Prog* **2006**, *22*, 141.

[13] H. Cölfen and L. M. Qi, *Chem Eur J* **2001**, *7* (1), 106.

[14] E. M. Pouget, P. H. H. Bomans, J. A. C. M. Goos, P. M. Frederik, G. de With and N. A. J. M. Sommerdijk, *Science* **2009**, *323* (5920), 1455.

[15] J. Rieger, T. Frechen, G. C. W. Heckmann, C. Schmidt and J. Thieme, *Faraday Discussions 136 on Nucleation and Crystal Growth*, Roy. Soc. Chem., London, 265–278.

[16] J. Bolze, B. Peng, N. Dingenouts, P. Panine, T. Narayanan and M. Ballauff, *Langmuir* **2002**, *18* (22), 8364.

[17] D. Pontoni, J. Bolze, N. Dingenouts, T. Narayanan and M. Ballauff, *J Phys Chem B* **2003**, *107* (22), 5123.

[18] T. Chen, A. Neville, K. Sorbie and Z. Zhong, *Faraday Discussions 136 on Nucleation and Crystal Growth*, Roy. Soc. Chem., London, 355–366.

[19] J. Rieger, J. Thieme and C. Schmidt, *Langmuir* **2000**, *16* (22), 8300.

[20] M. Faatz, F. Gröhn and G. Wegner, *Adv Mater* **2004**, *16* (12), 996.

[21] M. Faatz, *Kontrollierte Fällung von amorphem Calciumcarbonat durch homogene Carbonatfreisetzung*, Ph.D. thesis, Johannes Gutenberg – Universität, Mainz **2005**.

[22] V. K. LaMer and R. H. Dinegar, *J Am Chem Soc* **1950**, *72* (11), 4847.

[23] A. Navrotsky, *Proc Nat Acad Sci USA* **2004**, *101* (33), 12096.

[24] D. Horn and J. Rieger, *Angew Chem Int Ed* **2001**, *40*, 4340.

[25] H. Haberkorn, D. Franke, T. Frechen, W. Goesele and J. Rieger, *J Coll Interface Sci* **2003**, *259* (1), 112.

[26] A. Pimpinelli and J.Villain, *Physics of Crystal Growth*, Cambridge University Press, Cambridge, England **1998**, ISBN 0-521-55855-7.

[27] J. Leiterer, F. E. A. F. Thünemann and U. Panne, *Z Anorg Anal Chem* **2006**, *632* (12-13), 2132.

[28] E. G. Lierke, *Acustica* **1996**, *82* (2), 220.

[29] S. Santesson and S. Nilsson, *Anal Bioanal Chem* **2004**, *378* (7), 1704.

[30] D. M. Herlach, *Annual Review of Materials Science* **1991**, *21*, 23.

[31] D. M. Herlach, R. F. Cochrane, I. Egry, H. J. Fecht and A. L. Greer, *Int Mater Res* **1993**, *38* (6), 273.

[32] K. F. Kelton, G. W. Lee, A. K. Gangopadhyay, R. W. Hyers, T. J. Rathz, J. R. Rogers, M. B. Robinson and D. S. Robinson, *Phys Rev Lett* **2003**, *90* (19), 195504.

[33] H. P. Klug and L. E. Alexander, *X-Ray Diffraction Procedures: For Polycrystalline and Amorphous Materials*, Wiley–Interscience, Weinheim, Germany, 2nd ed. **1974**.

[34] H. Borchert, E. V. Shevehenko, A. Robert, I. Mekis, A. Kornowski, G. Grubel and H. Weller, *Langmuir* **2005**, *21* (5), 1931.

[35] H. M. Rietveld, *J Appl Cryst* **1969**, *2*, 65.

[36] R. D. Deegan, O. Bakajin, T. F.Dupont, G. Huber, S. R. Nagel and T. A.Witten, *Nature* **1997**, *389*, 827.

[37] N. Loges, K. Graf, L. Nasdala and W. Tremel, *Langmuir* **2006**, *22*, 3073.

[38] M. Avrami, *J Chem Phys* **1939**, *7* (12), 1103.

References

[39] M. Avrami, *J Chem Phys* **1940**, *8* (2), 212.

[40] M. Avrami, *J Chem Phys* **1941**, *9* (2), 177.

[41] A. R. West, *Solid State Chemistry and its Applications*, John Wiley & Sons, Chichester, England **1984**.

[42] W. Ostwald, *Z Phys Chem* **1897**, *22*, 289.

[43] Q. Shen, H. Wei, Y. Zhou, Y. Huang, H. Yang, D. Wang and D. Xu, *J Phys Chem B* **2006**, *110*, 2994.

[44] Y. Kitano, *Bull Chem Soc Japan* **1962**, *35* (12), 1973.

[45] L. B. Gower and D. J. Odom, *J Cryst Growth* **2000**, *210* (4), 719.

[46] E. J. W. Verwey and J. T. G. Overbeek, *Theory of the stability of lyophobic colloids. The interaction of sol particles having a electrical double layer.*, Elsevier Pub. Comp., Amsterdam, New York **1948**.

[47] D. Schwahn, M. Balz, M. Bartz, A. Fomenko and W. Tremel, *J Appl Cryst* **2003**, *36*, 583.

[48] D. Gittins and F. Caruso, *Angew Chemie Int Ed* **2001**, *113*, 568.

[49] O. Paris, C. H. Li, S. Siegel, G. Weseloh, F. Emmerling, H. Riesemeier, A. Erko and P. Fratzl, *J Appl Cryst* **2007**, *40*, 466.

[50] A. P. Hammersley, S. O. Svensson, M. Hanfland, A. N. Fitch and D. Hausermann, *High Press Res* **1996**, *14* (4-6), 235.

4 Nonclassical Homogenous Formation of Divalent Metal Carbonate Minerals

Abstract The homogenous formation of divalent metal carbonates from their bicarbonate solutions were studied by *in situ* X-ray scattering and transmission electron microscopy. It will be shown that the homogenous formation of metal carbonates (MCO_3, M = Ca, Sr, Ba, Mn, Cd, Pb) proceeds via an amorphous liquid-like intermediate in analogy to the formation of calcium carbonate. This amorphous transient phase is formed in the absence of any stabilizing polymer or additive at neutral pH. Beside a electrostatical stabilization, the stability of the liquid-like precursor may be related to the presence of various species such as carbonate, bicarbonate and non-dissociated carbonic acid involved in the process. The formation of an amorphous liquid-phase mineral precursor seems to be a characteristic of the homogenous formation of carbonates at neutral pH and indicates nonclassical homogenous crystallization of divalent metal carbonates.

Associated publications   *Nanoscale* **2011**, *3*, 1158–65.

Introduction

Because of its apparent simplicity, calcium carbonate is a popular model system for the study of nucleation and crystallization of minerals. Not long ago, the formation of calcium carbonate was discussed within the classical picture of crystallization based on the assumption that the formation of calcium carbonate crystals proceeds via nucleation and growth. However, during the past years there has been increasing evidence that amorphous calcium carbonate (ACC) plays a crucial role in crystallization. ACC is the most unstable form of calcium carbonate, and under ambient conditions it transforms quickly into more stable crystalline forms, such as vaterite and calcite.[1,2] Many mineralization processes are now believed to occur through the transformation of a transient amorphous precursor,[3] which acts as a reactive in intermediate in generating complex functional materials. Odom and Gower were the first to postulate the existence of a phase prior to the ACC phase called a polymer-induced liquid-precursor (PILP) phase,[4] which is a highly hydrated phase considered to be even more labile than the solid amorphous phase. Various analytical methods have been utilized to observe these initial formation steps, such as fast drying,[4] cryogenic transmission electron microscopy (cryo-TEM),[5,6] X-ray microscopy,[7] and small- and wide-angle X-ray scattering (SAXS, WAXS).[8–10] Rieger et al. studied the formation of calcium carbonate at high supersaturation ($c \approx 0.01$ mol l^{-1} during precipitation) after rapid mixing of the reactants $CaCl_2$ and Na_2CO_3.[6,7] Cryo-TEM studies revealed the formation of emulsion-like structures preceding the precursor stage and triggered speculations about a spinodal phase separation between a denser and a less dense phase. Faatz et al. reported the formation of calcium carbonate from a reaction of calcium chloride with carbon dioxide,[11,12] which was homogenously released to the solution through the alkaline hydrolysis of alkyl carbonate. Here the homogenous formation of CO_2 in the reaction medium prevents the formation of a gas/liquid interface, and the formation of amorphous calcium carbonate is postulated to proceed by a liquid-liquid binodal phase separation mechanism.[11] No analytical support for the formation of the proposed emulsion-like early stages was provided.

The precedent chapter concerning a transmission electron microscopy (TEM) study of the contact-free crystallization of calcium carbonate revealed the existence and homogenous formation of a liquid amorphous precursor phase in the absence of additives, which supports the idea of a nonclassical route for calcium carbonate formation (*cf.* Chp. 3). In order to achieve a large homogenous supersaturation of the solution and to suppress nucleation by the action of foreign bodies (*e. g.* macromolecules, spectator ions, liquid/liquid- or solid/liquid-interfaces, like from vessel walls or due to mixing processes) the crystallization experiments starting from calcium bicarbonate solution were carried out in an ultrasonic levitator. TEM revealed that a liquid calcium carbonate phase was formed, which solidifies in the course of time. Cryogenic scanning electron microscopy (cryo-SEM) proved that the observed liquid particles form homogenously within the droplet and not by heterogenous nucleation at the air/water interface.

In this chapter, it shall be pointed out with the aid of TEM and SAXS studies that nonclassical crystallization via a liquid precursor phase applies not only to calcium carbonate ($CaCO_3$) but also to other divalent carbonate minerals (MCO_3, M = Ca, Sr Ba, Mn, Cd, Pb).

Experimental Part Saturated bicarbonate solutions were prepared analogously as before by treating a slurry of the respective carbonate (MCO_3, M = Ca, Sr, Ba, Mn, Cd or Pb) in ultrapure water (Millipore Synergy 185 with UV photo oxidation, $18.2\,M\Omega\,cm^{-1}$) with carbon dioxide (Westfalen AG). The slurry was filtered down to 20 nm with a cascade of syringe filters (0.1 µm Millipore

Alkaline Earth Carbonates	Life-span	Non Alkaline Earth Carbonates	Life-span
$CaCO_3$	400 s	$MnCO_3$	400 s
$BaCO_3$	300 s	$PbCO_3$	400 s
$SrCO_3$	250 s	$CdCO_3$	800 s

Table 4.1 Maximum life-span of liquid intermediates of the six investigated carbonate minerals as roughly estimated by TEM analysis.

Millex VV and 20 nm Milipore Anotop) and treated again with carbon dioxide for at least 1 h per 10 ml filtrate. The resulting bicarbonate solution is virtually free of foreign and metal carbonate nuclei as verified by SAXS and TEM. The mineralization of a 4 µL droplet of mother solution was performed contact-free by means of an ultrasonic levitator (58 kHz, Tec 5, Oberursel, Germany). TEM investigations were carried out with a Phillips EM 420 running at 120 kV, equipped with an ORCA-ER Camera (1024 × 1024 pixel) and run with AMT Image Capture Engine v5.42.540a. TEM samples were prepared by transferring a droplet to a lacey-coated TEM grid (Plano, Germany) and air-drying. A washing step was omitted because this lead to complete dissolution of the liquid transient phase.

Wide-angle and small-angle X-ray scattering (WAXS, SAXS) experiments were performed at the µSpot beamline of BESSY, which provides a monochromatic beam ($\lambda = 1.00257$ Å based on calibration with corundum) and a photon flux of about $10^9 \, s^{-1}$.[13] Due to a small beam diameter of 20 µm a mathematical desmearing of the experimental scattering intensity function was obsolete. Reduction of scattering data was accomplished with the data analysis program FIT2D.[14]

Liquid-like particles were obtained for all samples as depicted in Fig. 4.1. The low contrast variation within the particles indicates their liquid character. Solid spherical particles would show a distinct increase in contrast from the surface to the center of the particles. The existence of liquid state could be proven with the aid of an experimental artefact. When preparing the specimens on a TEM grid, excess mother solution had to be removed. The solvent removal occasionally exerted a flow on the settled fluid mineral particles (see Fig. 4.3) and lends further evidence for their liquid state.

The life-span of the emulsion state was be roughly assessed by TEM analysis after different periods of levitation (see Table 4.1). For all six compounds, crystallization of the corresponding metal carbonate was induced in the electron microscope after exposing the samples containing liquid-like particles to the electron beam. The crystallization may be attributed to a loss of water of hydration.

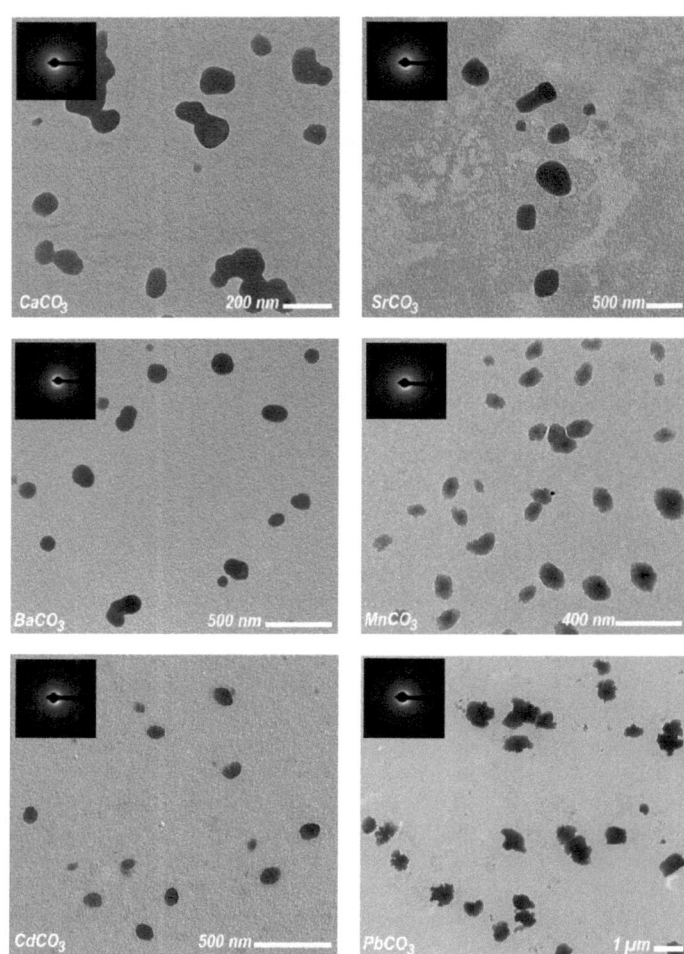

Figure 4.1 Transmission electron micrographs of liquid-like particles formed by metal carbonates (MCO_3, M = Ca, Sr, Ba, Mn, Cd, Pb). The respective electron diffraction is shown in the inset, which indicate the amorphous state of the particles.

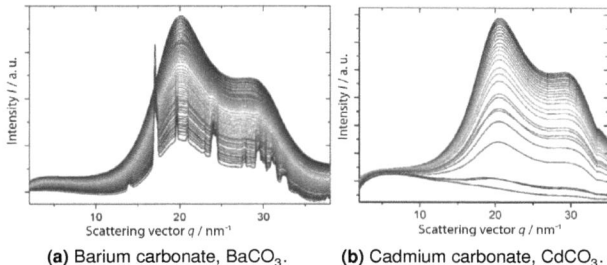

(a) Barium carbonate, BaCO$_3$.   **(b)** Cadmium carbonate, CdCO$_3$.

Figure 4.2 Wide-angle scattering monitored during evaporation of a saturated divalent bicarbonate solution under levitated contact-free conditions, here **(a)** barium carbonate and **(b)** cadmium carbonate. The first scattering curves are colored in blue; the scattering during the last state of experiment is printed in red. Barium carbonate finally forms the mineral phase witherite, which is orthorhombic and isotypic with aragonite (high pressure phase of CaCO$_3$) and stable at ambient conditions. Cadmium bicarbonate does not end up in crystalline material, cadmium carbonate remains amorphous.

The formation of the crystalline carbonate phases from the liquid intermediate was monitored *in situ* by wide-angle scattering (WAXS, see Fig. 4.2). The first patterns only show diffuse scattering of water, which vanishes gradually as the water evaporates. The Bragg reflections of the mineral appear according to their intensity, *i. e.* the (104) reflection of calcite for CaCO$_3$, or the (111) reflection of strontianite (SrCO$_3$) and witherite (BaCO$_3$) as the first. Cadmium carbonate does not develop a crystalline mineral phase in detectable amounts during the levitation process until the evaporation solvent evaporation is complete. This is compatible with the long life-span of the liquid amorphous intermediate of cadmium carbonate (CdCO$_3$).

The occurrence of liquid precursors in six different carbonate minerals demonstrates that the nonclassical crystallization route via a liquid intermediate is not a singular phenomenon of the extensively studied calcium carbonate system. It seems to be a characteristic of divalent metal carbonate compounds and may be related to the degrees of freedom of carbonate anions (*e. g.* rotational, tilting, protonation) that pose a potential barrier for crystallization. No additives were needed to stabilize the liquid precursors.

Why is a liquid mineral emulsion with a lifetime of several minutes formed? Besides the electrostatic stabilization, which was proposed in the preceding chapter, the liquid intermediate gains its stability from a 3D-disordered network built up by a variety of interacting species. In a bicarbonate buffer solution at pH $\approx$ 7, numerous carbonate species coexist, *i. e.* the CO$_3^{2-}$, HCO$_3^-$ and undissociated H$_2$CO$_3$.[15–17] These carbonate species may serve

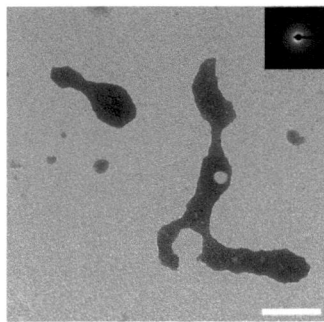

Figure 4.3 Precursor particles of the liquid-like barium carbonate mineral phase which experienced a forced flow. The particle shape illustrates its fluid character. Scale bar: 500 nm.

as mono- and/or bidentate ligands for the divalent metal ions. Uncoordinated carbonate species and all possible varieties of metal complexes can build up a hydrated network based on hydrogen bonding. The presence of *(i)* various bonding partners, *(ii)* variable coordination geometries and ligands of the metal center, *(iii)* alternating coordination numbers of the carbonate groups and *(iv)* the associated variation of local structures favors the formation of the hydrated non-crystalline phase. In the bicarbonate buffer regime the variation of local structures will develop a maximum as all three different carbonate species are present in comparable concentrations. At higher pH, the CO_3^{2-} anion will dominate and the number of carbonate complexes will decrease accordingly. Consequently, a less hydrated and less disordered network gets formed, which can transform to a crystalline material more easily as it resembles one of the carbonate mineral crystal structures more closely.[18] Prenucleation clusters, which were recently detected in carbonate buffer solution at pH 9–10 and were roughly estimated to consist of 70 calcium and 70 carbonate ions, may be viewed as a representative of such a hydrated polynuclear coordination network.[17] The size distribution (2–4 nm) of the prenucleation clusters is determined by the electrostatic stabilization and their synthesis at undersaturated concentration. The surface charge density is a function of the ion activity and the pH, *i. e.* it critically depends on the experimental conditions: The higher the pH and thus the surface charge density, the smaller will be the stable particle radius.

Conclusions In conclusion, it could be shown that the homogenous formation of metal carbonates (MCO_3, M = Ca, Sr, Ba, Mn, Cd, Pb) proceeds via an amorphous liquid-like state. This amorphous phase is formed in the absence of any stabilizing polymers or additives at neutral pH, and the transient mineral emulsion is stabilized electrostatically. Its stability may be related to the presence of various species such as carbonate, bicarbonate and non-dissociated carbonic acid involved in the process. It is not limited to calcium carbonate and seems to be a quite general phenomenon. The formation of an amorphous liquid-phase mineral precursor seems to be a characteristic of the homogenous formation of carbonates. It is not limited to calcium carbonate and seems to be a quite general phenomenon. All presented observations suggest that the present ideas of calcium carbonate morphosynthesis could be applicable for a wider range of minerals if a liquid precursor phase plays a key role during morphogenesis, *e. g.* like in case of strontianite and vaterite nanowires,[19,20] the PILP process,[4] or meso- and nonclassical crystallizations in general.[21,22]

References

[1] L. Addadi, S. Raz and S. Weiner, *Adv Mater* **2003**, *15*, 959.
[2] F. Meldrum, *Int Mater Rev* **2003**, *48*, 187.
[3] E. Beniash, L. Addadi and S. Weiner, *J Struct Biol* **1999**, *125* (1), 50.
[4] L. B. Gower and D. J. Odom, *J Cryst Growth* **2000**, *210* (4), 719.
[5] E. M. Pouget, P. H. H. Bomans, J. A. C. M. Goos, P. M. Frederik, G. de With and N. A. J. M. Sommerdijk, *Science* **2009**, *323* (5920), 1455.
[6] J. Rieger, T. Frechen, G. C. W. Heckmann, C. Schmidt and J. Thieme, *Faraday Discussions 136 on Nucleation and Crystal Growth*, Roy. Soc. Chem., London, 265–278.
[7] J. Rieger, J. Thieme and C. Schmidt, *Langmuir* **2000**, *16* (22), 8300.
[8] J. Bolze, B. Peng, N. Dingenouts, P. Panine, T. Narayanan and M. Ballauff, *Langmuir* **2002**, *18* (22), 8364.
[9] D. Pontoni, J. Bolze, N. Dingenouts, T. Narayanan and M. Ballauff, *J Phys Chem B* **2003**, *107* (22), 5123.
[10] T. Chen, A. Neville, K. Sorbie and Z. Zhong, *Faraday Discussions 136 on Nucleation and Crystal Growth*, Roy. Soc. Chem., London, 355–366.
[11] M. Faatz, F. Gröhn and G. Wegner, *Adv Mater* **2004**, *16* (12), 996.
[12] M. Faatz, F. Gröhn and G. Wegner, *Mater Sci Eng C* **2005**, *25* (2), 153.
[13] O. Paris, C. H. Li, S. Siegel, G. Weseloh, F. Emmerling, H. Riesemeier, A. Erko and P. Fratzl, *J Appl Cryst* **2007**, *40*, 466.
[14] A. P. Hammersley, S. O. Svensson, M. Hanfland, A. N. Fitch and D. Hausermann, *High Press Res* **1996**, *14* (4-6), 235.
[15] D. D. Tommaso and N. H. de Leeuw, *J Phys Chem B* **2008**, *112*, 6965.
[16] N. Loges, S. E. Wolf, M. Panthöfer, L. Müller, M.-C. Reinnig, T. Hoffmann and W. Tremel, *Angew Chem Int Ed* **2008**, *120* (20), 3741.

References

[17] D. Gebauer, A. Völkel and H. Cölfen, *Science* **2009**, *322*, 1819.

[18] R. S. K. Lam, J. M. Charnock, A. Lennie and F. C. Meldrum, *CrystEngComm* **2007**, *9*, 1226.

[19] M. Balz, H. Therese, M. Kappl, L. Nasdala, W. Hofmeister, H.-J. Butt and W. Tremel, *Langmuir* **2005**, *21* (9), 3981.

[20] M. Balz, H. Therese, J. Li, J. S. Gutmann, M. Kappl, L. Nasdala, W. Hofmeister, H.-J. Butt and W. Tremel, *Adv Funct Mater* **2005**, *15* (4), 683.

[21] H. Cölfen and M. Antonietti, *Mesocrystals and Nonclassical Crystallization*, John Wiley & Sons Ltd, The Atrium, Southern Gate, Chichester **2008**, ISBN 987-0-470-02981-7.

[22] D. Volkmer, M. Harms, L. A. Gower and A. Ziegler, *Angew Chemie Int Ed* **2005**, *44*, 639.

5 Stabilization and Destabilization of the Transient Liquid Calcium Carbonate Precursor Phase by Ovo-Proteins

Abstract The impact of the ovo-proteins ovalbumin and lysozyme on homogenous formation of the liquid-like calcium carbonate precursor was studied. If the differently charged biopolymers ovalbumin and lysozyme are present at $7.5\,\mathrm{g\,l^{-1}}$, the basic and positively charged protein lysozyme ($pI = 9.3$) destabilizes the emulsified state whereas the negatively charged acidic protein ovalbumin ($pI = 4.7$) extends its life-span. As lysozyme demulsifies the transient state, the assumption of negative surface charge and an electrostatic stabilization of the pure emulsified liquid calcium carbonate precursor is reasonable. The 'polymer-induced liquid-precursor', reported by L. B. Gower, rather seems to be affected by polymers like poly-acrylic acid or ovalbumin in terms of depletion destabilization than literally induced by acidic proteins and polymers.

Ovalbumin represents the first natural protein, which behaves commensurable to the PILP model. In the light of the presented data it can be further speculated, whether ovalbumin takes a key role during the first stages of egg shell formation. Ovalbumin is capable to serve *(a)* as effective stabilization agent for a transient mineral precursor, *(b)* as a storage protein (aggregate) of inorganic egg shell components and *(c)* as a prevention of undirected mineralization in favor of a directed mineralization of the egg shell.

Associated publications   *J Am Chem Soc* **2011**, submitted.
J Am Chem Soc **2008**, *190* (21), 6879–92.

Introduction

Gaining control over morphogenesis and phase selection emerges today as a pivotal challenge for material sciences and technology since the entire characteristics of a material can be altered by modifying its shape, phase size, and substructure. Biominerals show prominently, that very simple and abundant minerals may exhibit superior properties in order to serve as sensors, skeletal support, or protection of soft tissues.[1-3] Combining inorganic toughness with organic elasticity, biominerals excel their purely inorganic counterparts. Intra-crystalline biomacromolecules (*e. g.* proteins, glycoproteins, polysaccharides, or proteoglycans) are often highly acidic and are assumed to induce and control nucleation, phase, growth, size and shape of the emerging biomineral.[4-6] In case of the calcified layer of avian eggshells, the mineral layer consists of roughly 5% organic material beside the mineral phase; the latter is almost exclusively the thermodynamic stable mineral phase calcite ($CaCO_3$).[7] The avian eggshell is one of the fast forming biominerals; roughly 5 g of calcium carbonate is deposited during 22 h of egg shell formation in an acellular environment. Nevertheless, the eggshell is crystallographic highly orientated, permeable at the same time and its breaking is crystallographic controlled.[1,8,9] Ovalbumin and lysozyme are two of the egg white proteins which are present during the initial stage of egg shell formation.[10,11] The acidic glycoprotein ovalbumin ($pI = 4.7$, 45 kDa) is the most dominant protein during the initial stage, and contributes 54% to the hen egg white but its biological function is still unclear.[12-14] Lysozyme ($pI = 9.3$, 14.3 kDa, 3.5% of the hen egg white) fulfills antibacterial tasks and thus plays a chemical protective function during avian embryonic development.[7,15] Different observations strongly support the idea that the eggshell matrix compounds regulate the egg shell formation. Each phase of shell mineralization (nucleation, rapid crystal growth and the completion of shell formation) is associated with a specific distribution of biological macromolecules in the uterine fluid.[16] Furthermore, bicarbonate concentrations are close to 100 mmol l^{-1} in the uterine fluid, and the calcium concentration ranges from 5 to 10 mmol l^{-1}. The inorganic constituents are 60- to 100-fold hypersaturated with regard to the solubility product of calcite. Therefore, the components of the uterine fluid have to prevent unfocused precipitation in favor of a controlled, spatially restricted growth of the egg shell via a precursor or a storage process. Today, the amorphous phase of calcium carbonate (ACC) is assumed to take this role of a transient precursor phase during several biomineralization and biomimetic processes,[17-19] as it was shown exemplarily in case of larvae of sea urchins or molluscan bivalves.[20,21] ACC is the most unstable form of calcium carbonate, and under ambient conditions, it transforms quickly into more stable crystalline forms, like vaterite and calcite.[5] However, Gower suggested the existence of a liquid transient phase prior to ACC which formation is induced by the addition of small poly-anionic polymers like polyaspartate or polyacrylate. These polymers are assumed to gather and to sequester cations and, as a consequence, to attenuate the supersaturation and to delay the crystallization.[22] Meanwhile, this polymer-induced liquid-precursor (PILP) phase, which is considered to be even more labile than the solid amorphous phase, was repeatedly employed to prepare different exceptional morphologies of calcium carbonate.[22–26]

In the preceding chapter (Chp. 3) the existence of a transient and liquid-like amorphous calcium carbonate phase was discussed, which is formed without an induction by small anionic polymers at the outset of the precipitation of calcium carbonate. By employing a contact-free sample environment, heterogenous nucleation was suppressed and cryogenic scanning electron microscopy (cryo-SEM) proved this approach to be successful, as the observed liquid particles form homogenously within the droplet and not by heterogenous nucleation at the air/water interface. This chapter addresses the issue, how ovo-proteins of different type and isoelectric point (pI), which are present at the first stage of egg shell formation, bias the formation of calcium carbonate in general and how these natural proteins influence the liquid precursor of amorphous calcium carbonate in detail. The attendant proteins have to prevent undirected mineralization and should provide support during directed mineralization, maybe in terms of a storage function or stabilization for a transient precursor. As a model, the acidic protein ovalbumin is contrasted ($pI = 4.7$) with the basic protein lysozyme ($pI = 9.3$) which both are present during the first stage of egg shell formation. The crystallization was carried out homogenously according to the Kitano method,[27] where calcium carbonate is formed by slow evaporation of water and concomitant slow release of carbon dioxide from a saturated solution of calcium bicarbonate. This precipitation proceeds slowly at nearly neutral pH because of the inherent bicarbonate puffer system ($p\mathrm{H} = 7.35 - 7.45$). In order to study the impact of proteins on the homogenous formation of calcium carbonate, an ultrasonic levitator was employed to prevent heterogenous influences of foreign materials and their phase boundaries as described previously (*cf.* Chp. 3).[28] *In situ* X-ray scattering experiments were performed at a synchrotron microspot beamline to monitor the mineral phase formation. In addition, different stages of the crystallization were characterized by transmission and scanning electron microscopy (TEM, SEM).

One droplet with a volume of 4 µL of solution was levitated, which contained beside the respective ovo-proteins (7.5 g l^{-1}), calcium bicarbonate at a saturated concentration (≈ 10 mmol l^{-1}). Due to slow evaporation of water and concomitant slow release of carbon dioxide calcium carbonate was formed. The high concentration of proteins should resemble the conditions in the uterine fluid. The precipitation was followed by means of *in situ* wide-angle X-ray scattering; the respective patterns are shown in Fig. 5.1. The appearance of Bragg reflexes indicated the incipient crystallization—more strictly, the formation of the first crystalline phase as amorphous phases may precede it, which are undetectable by diffraction. The time which passed until these first reflexes appeared was quantified by an Avrami analysis of the integral reflex intensities.[29–31] Corrected and normalized integral intensities of the (104) reflex were fitted to a Weibull function, whose respective inflexion point was used as a comparative value t_p^{104} which corresponds to the elapsed time when calcite formation ceased.

In absence of proteins, the first detectable reflex belonged to the {104} set of calcite lattice planes. Later, other reflections followed: (104), (102), (110), (113), (202) appeared almost synchronous whereas the weakest reflection (006) was detected as the latest (Fig. 5.1a). A t_p^{104} value of 34.5 min was estimated (Fig. 5.2a). Evaluating the final WAXS pattern of the dry sample, only the stable calcite phase was found (cf. Fig. 5.2b). The dry sample was investigated with scanning electron microscopy and showed that spherical solid particles were present along with rhombohedral calcite crystals (Fig. 5.3a); the former particles consisted of solidified dry amorphous calcium carbonate, which did not transform into crystalline material (cf. Chp. 3).[28] In presence of lysozyme, the evolution of the WAXS patterns did not change significantly (Fig. 5.1b). Slow increases in intensity in the small angle regime indicated agglomeration of lysozyme;[32,33] these signals disappeared when the sample had reached its dry state. The kinetics of the precipitation process did not differ significantly from those of a protein-free mineralization ($t_p^{104} = 31$ min, cf. Fig. 5.2a). As found in absence of proteins, the final mineral phase is the thermodynamic stable phase calcite. Presence of ovalbumin affected the precipitation of calcium carbonate very considerably. The formation of crystalline calcium carbonate was strongly retarded; after 53 min reflexes appear quite abruptly ($t_p^{104} = 53$ min, cf. Fig. 5.2a). In contrast to the two other experiments with lysozyme and without protein, only the presence of ovalbumin led to a mixture of calcium carbonate phases (Fig. 5.2b). A ratio of 80.16% vaterite vs. 19.83% calcite was determined based on Rietveld refinements.[34] Small-angle scattering emerged later than in case of lysozyme. In a recent small-angle neutron-scattering (SANS) study,[35] ovalbumin was shown to collect a high load of calcium ions from solution. As a consequence of the concomitant breaking of intermolecular hydrogen bonds due to complexation of calcium by the protein's acidic groups, ovalbumin aggregates and refolds to form a Gaussian chain with large segments. TEM and standard microscopy revealed that these protein chains aggregate further to form fibrils featuring a high aspect ratio (cf. Fig. 5.4). In the final dry sample, these protein fibrils fully dominated the appearance of the residue (cf. Fig. 5.3). EDX spectroscopy revealed that the prominent knobs at the residue's outer surface consisted of calcium carbonate, whereas the other remains were protein aggregates.

As amorphous intermediates cannot be traced by diffraction, the early stages of mineralization were investigated by transmission electron microscopy. In

Results and Discussion

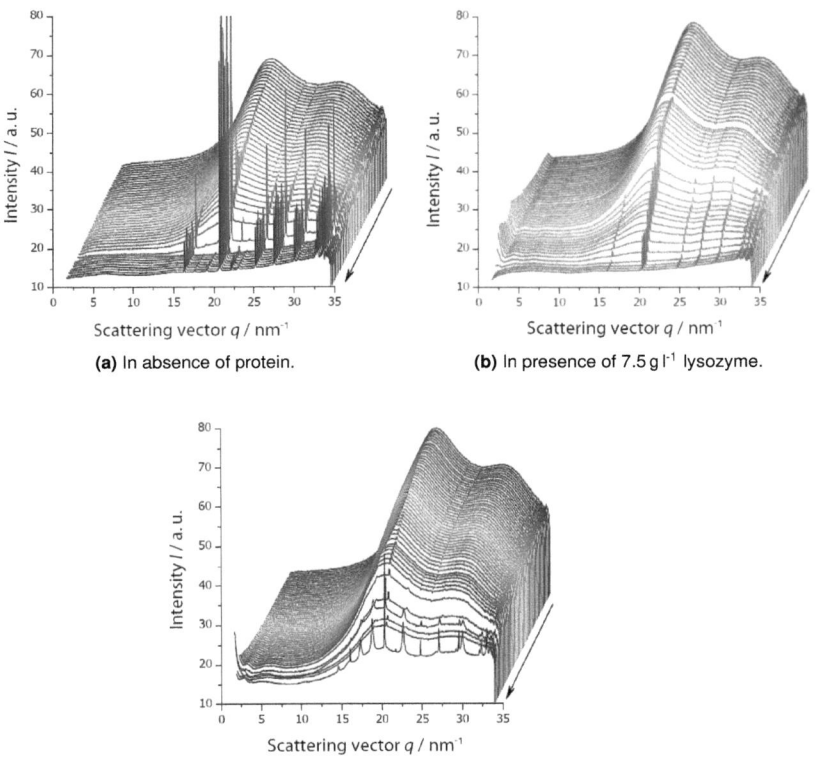

Figure 5.1 Evolution of scattering intensities during *in situ* monitoring of the evaporation of a levitated calcium bicarbonate solution in presence and absence of proteins.

45

5. Stabilization and Destabilization of the Liquid Precursor Phase by Ovo-Proteins

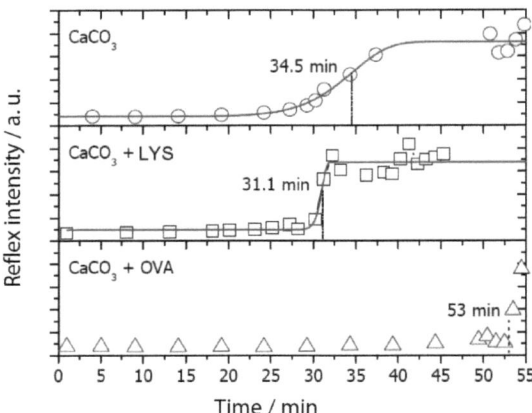

(a) Integral Intensity of the (104) reflexes as a function of time. The corresponding t_p^{104} values are marked by anchor lines.

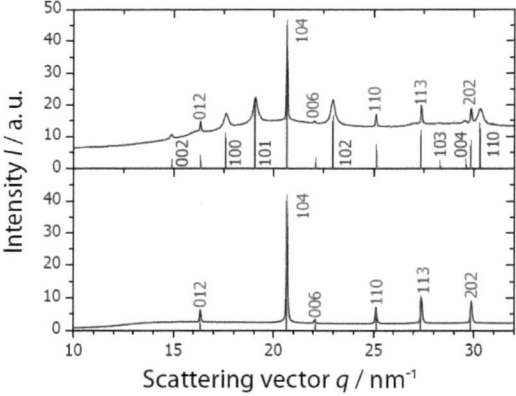

(b) WAXS patterns of the final stages of crystallization (black line) and their corresponding Miller indices (anchor lines). The lower pattern was obtained from protein-free precipitation and shows only Bragg reflexes of calcite (red anchor lines). The upper pattern of crystallization in presence of ovalbumin features Bragg reflexes both calcite (red anchor lines) and vaterite (blue anchor lines).

Figure 5.2 Integral Intensity of the (104) reflexes as a function of time and WAXS pattern of the final stages of crystallization.

 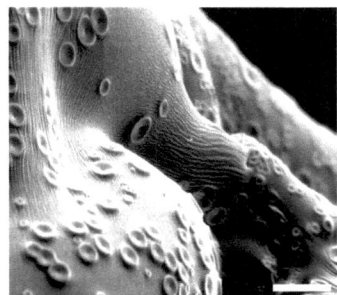

(a) Scale bar: 10 µm.   (b) Scale bar: 20 µm.

Figure 5.3 Scanning electron micrographs of the final stages of precipitation of (a) pure calcium carbonate, (b) calcium carbonate in presence of 7.5 g l^{-1} ovalbumin.

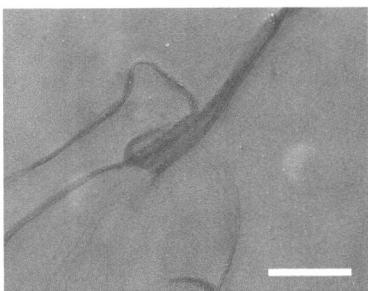 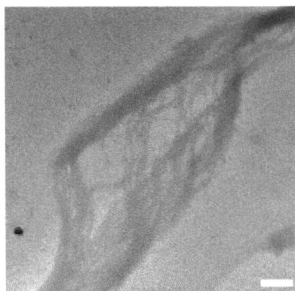

(a) Stained with bromthymol-blue.   (b) Scale bar: 500 nm.
Scale bar: 50 µm.

Figure 5.4 (a) Standard microscopy and (b) transmission electron micrographs of ovalbumin fibrils, whose formation was induced by the addition of 10 mmol l^{-1} calcium chloride.

case of the pure calcium bicarbonate solution, a liquid/liquid phase separation occurs at the outset of the precipitation prior formation of the crystalline calcite phase as shown in the previous chapter (*cf.* Fig. 5.5a and Chp. 3). The amorphous state of the droplets, which were formed during the early stages of precipitation, was ascertained by electron diffraction (ED, see inset in Fig. 5.5a). The low contrast variation of the droplets gave evidence of their liquid-like character; solid spherical particles would show a distinct increase in contrast from the particle boundary to their center whereas liquid droplets will undergo diffluence to flat particles if they settle down on a flat surface (here a coated TEM grid) and thus the contrast gradient due to a varying diameter of the particle would disappear.[28,36] Electrostatical stabilization of this emulsion-like state is reasonable as no other of the classical colloidal stabilization mechanisms—*i. e.* sterical or depletion stabilization— can apply. The droplets consisted of highly hydrated calcium carbonate; the radiative stress during the TEM analysis induced crystallization of calcium carbonate due to loss of water of hydration. In presence of lysozyme during the precipitation process, the salient emulsion-like appearance of the precursor vanished. Instead of individually droplets, an intense coalescence seemed to have occurred during the first 400 s (*cf.* Fig. 5.5b). Electron diffraction (ED) showed these structures to be as well non-crystalline (inset of Fig. 5.5b). As implied by the deviation of the WAXS patterns, the impact of ovalbumin on the precursor structures differed distinctly from the effect lysozyme induced. The liquid intermediate was considerably stabilized and its life-span was greatly extended in presence of ovalbumin during the precipitation. Whereas sporadic crystalline material could be found after $\approx 500\,\text{s}$ by TEM in absence of proteins, an emulsion-like state still existed after 500 s (Fig. 5.5c) and persisted up to remarkable 1 000 s (*cf.* Fig. 5.5d). The appearance of the ovalbumin-stabilized mineral emulsion completely resembled the one which formed in absence of protein (*cf.* Fig. 5.5a *vs.* 5.5d) and the amorphous state of the droplets could be verified by ED. The question arose, whether ovalbumin was accumulated in the calcium carbonate droplets or if it remained in the mother solution. Samples, which were obtained after 300 s at a protein concentration of $0.5\,\text{g}\,\text{l}^{-1}$, were fixated by drying at 40° C for 48 h and studied by a polyclonal immunogold (IG) labeling in order to locate ovalbumin. As shown in Fig. 5.6, the droplet suffered from the steps of rinsing, which are inevitable during IG-labeling. Nevertheless, IG-labeling could be accomplished and the labeling was found to be associated partially with the droplets and—to a much lesser extend—randomly distributed as well. A labeling ratio of 1 : 6.78 of unbound *vs.* droplet-associated labeling was determined. Thus, ovalbumin remains partially in solution, but a moiety is in fact incorporated in the droplets of liquid calcium carbonate.

Results and Discussion

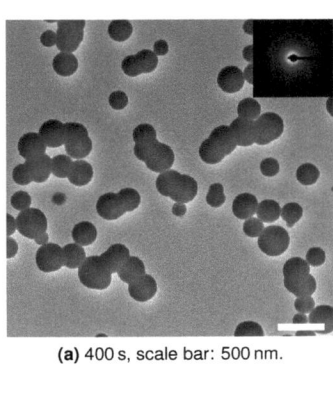

(a) 400 s, scale bar: 500 nm.

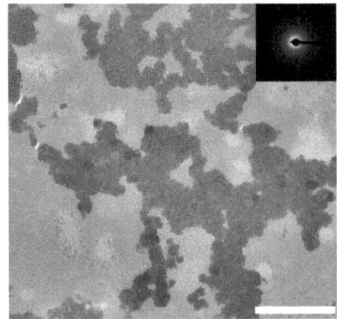

(b) 400 s, scale bar: 500 nm.

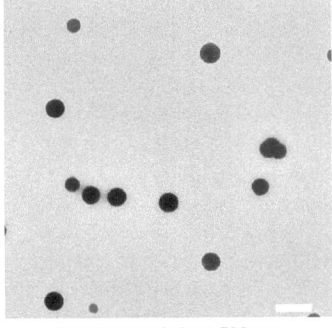

(c) 500 s, scale bar: 500 nm.

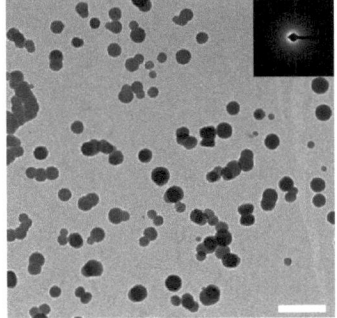

(d) 1 000 s, scale bar: 500 nm.

Figure 5.5 Transmission electron micrographs of precipitations of calcium carbonate **(a)** in absence of ovo-proteins, **(b)** in presence of 7.5 g l^{-1} lysozyme, **(c, d)** in presence of 7.5 g l^{-1} ovalbumin. The sampling was done at different reaction times.

5. *Stabilization and Destabilization of the Liquid Precursor Phase by Ovo-Proteins*

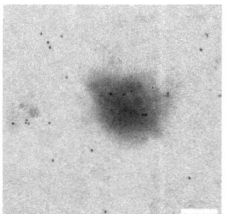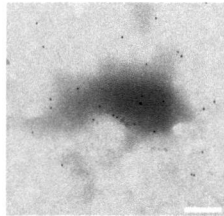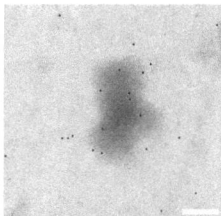

Figure 5.6 Transmission electron micrographs of samples, which were obtained after 300 s at a protein concentration of 0.5 g l^{-1}. Polyclonal immunogold labeling indicates that ovalbumin is partially incorporated in the liquid amorphous calcium carbonate precursor. Scale bars: 200 nm.

Conclusions   Ovalbumin is capable to stabilize extensively the liquid precursor phase of calcium carbonate, whereas lysozyme rather destabilizes the transient phase. According to WAXS, this induces a shift in phase composition of the later calcium carbonate minerals. According to the preceding chapter (Chp. 3), the presence of the liquid ACC precursor is based on formation of an extensive disordered network composed of hydrogen bridges of numerous species (water, carbonate, bicarbonate, non-dissociated carbonic acid and different hydrated calcium complexes). The occurrence of a liquid amorphous state during calcium carbonate formation is thus rather a specialty of calcium carbonate formation then induced by anionic polymers. An electrostatical stabilization mechanism of the emulsified precursor in absence of polymers or proteins is reasonable, as no other colloidal stabilization mechanisms can apply.

If differently charged biopolymers are present in such an emulsion at 7.5 g l^{-1}, the basic and positively charged protein lysozyme ($pI = 9.3$) destabilizes the emulsified state whereas the negatively charged acidic protein ovalbumin ($pI = 4.7$) extends its life-span. It is a well-known behavior of colloidal systems that if the additive is of opposite charge in respect to the emulsified moiety phase, it will destabilize the emulsion either by compensating the surface charges or interconnecting the emulsified droplets.[37] Based on the presented findings, one can deduce that the emulsified liquid calcium carbonate precursor is actually negatively charged and the positively charged lysozyme induces flocculation via charge neutralisation or interconnection. As a further consequence, electrostatical stabilization of the pure, additive-free mineral emulsion is reasonable as well.

The influences which ovalbumin exerts on the liquid mineral phase are multifold. Recent data from small-angle neutron-scattering showed ovalbumin to act like a 'cation sponge';[35] it accumulates calcium ions from solution which are complexated by the protein's carboxylic groups. In other words, it decreases dramatically the calcium activity of the bulk solution to locally increase the calcium concentration next to the protein. The decrease in calcium activity due to complexation relieves the solution's supersaturation and thus decelerates formation of crystalline material whose formation would require higher supersaturation based on their solubility products, and thus extends the life-time of the liquid and amorphous precursors. From a colloidal chemical point of view, one may call this depletion stabilization. The process of depletion stabilization is invariably preceded by depletion flocculation. Thus, it is reasonable that the original PILP coating effect, which was reported by Gower *et al.* at much lower polymer concentrations,[22-24] actually is a result of a demulsifying process based on depletion destabilization.

Ovalbumin, found by SANS to be loaded with calcium, represents a 'fluctuation in calcium concentration'. Such concentration fluctuations are the crucial point during formation of a new phase (*cf.* Appendix Sec. 8). In pure solutions, these fluctuations are due to statistical processes (and thus seldom occurring). In the present case they are provided due to the presence of ovalbumin. Thus, ovalbumin promotes nucleation by gathering calcium ions and lowers the main activation barrier of phase separation; one may speculate whether this represents a switch from a spinodal to a binodal liquid/liquid phase separation process.[36,38] This behavior is perfectly commensurable with Gower's PILP concept extended by the fact that the formation of the liquid mineral precursor is not induced but rather facilitated and accelerated as it occurs as well in absence of polymers. In the later state of mineralization, the incorporated ovalbumin—or small anionic polymers in general—in the amorphous mineral phase increases the disorder of the 'glassy' state by providing an additional multitude of different binding and bridging possibilities and stabilizes thereby the amorphous state. The transformation to a crystalline state remains hampered and may yield in patterned crystalline material, in which the incorporated polymer is occluded in transition bars.[22,39]

5. Stabilization and Destabilization of the Liquid Precursor Phase by Ovo-Proteins

Summary   In summary, ovalbumin was shown to stabilize the liquid calcium carbonate phase remarkably under neutral and diffusion-controlled conditions and thus represents the first natural protein, which behaves commensurable to Gower's PILP process. As lysozyme demulsifies the transient state, the assumption of negative surface charge and an electrostatic stabilization of the pure emulsified liquid calcium carbonate precursor is reasonable. Concerning the PILP model, the liquid calcium carbonate phase rather seems to be affected by polymers in terms of depletion stabilization resp. destabilization than literally induced and thus triggered by acidic proteins and polymers during a PILP process. The liquid calcium carbonate phase can be regarded a specialty of additive-free calcium carbonate formation at neutral pH and forms in absence of polymers as well.

In the light of the presented data, ovalbumin seems to take a key role during egg shell formation. Ovalbumin can be speculated to serve as *(a)* effective stabilization agent for a transient mineral precursor, *(b)* a storage protein (aggregate) of the inorganic egg shell components and as a *(c)* prevention of undirected mineralization in favor of a directed mineralization of the egg shell.

Experimental Part

A suspension of $CaCO_3$ (*p. a.*, Sigma Aldrich) in ultrapure water (Millipore Synergy 185 with UV photo oxidation, $18.2\,M\Omega\,cm^{-1}$) was treated with carbon dioxide (Westfalen AG). The obtained saturated solution of calcium bicarbonate was filtered with a cascade of syringe filters, which consists of a 0.1 µm Millipore Millex VV followed by 20 nm Milipore Anotop in series. Afterwards, the filtered solution was treated again extensively with carbon dioxide to dissolve nuclei with diameters < 20 nm. Immediately after dissolving the desired protein (ovalbumin Grade V and lysozyme from chicken egg white; Sigma Aldrich) in the saturated solution of calcium bicarbonate at the requested concentration, one droplet of the solution with a volume of approximately 4 µL was manually injected in the ultrasonic levitator (Tec 5, Oberursel, Germany). The reaction was monitored by wide-angle X-ray scattering measurements (WAXS) performed at the µSpot beamline at BESSY, which is equipped with a double crystal monochromator, yielding a monochromatic ($\lambda = 1.00257\,\text{Å}$) beam. A pinhole system provides a beam of 20 µm cross section with a photon flux of about 10^9 per second and a ring current of 100 mA. Further information concerning the general setup are given by Paris *et al.*[40] Scattering patterns from corundum served as external calibration standard. No mathematical 'desmearing' of the experimental scattering intensity function was needed due to the small beam diameter of the incident beam. The levitated sample was monitored and remotely controlled during the whole experiment. The data was processed and converted to diagrams of scattering intensities I versus q by employing algorithms of the computer program FIT2D.[41] Avrami analysis was conducted with OriginPro 8, fitting the background-corrected and normalized integral intensity of the (104) reflex to the Weibull function $I = a - (a - b)\,e^{-(kt)^d}$, where t is the levitation time, and a, b, k and d are free parameters. TEM investigations were carried with a Phillips EM 420 running at 120 kV, equipped with an ORCA-ER Camera (1024 × 1024 pixel) and run with AMT Image Capture Engine v5.42.540a. The samples were prepared by transferring the respective droplet on a lacey-coated TEM grid (Plano, Germany), followed by washing with water and air-drying. SEM investigations were performed with a Zeiss DSM 940 running at 10 kV. For immunogold labeling, a classical labeling protocol was followed as described by Harris.[42] The samples were prepared by transferring the respective droplet on a lacey-coated TEM grid (Plano, Germany), followed by washing with water and fixating at 48° C for 48 h. The sample was blocked with 10% purified casein protein (Roche, Germany) and then incubated with a 1:1000 dilution of the primary antibody (IgG fraction of anti-ovalbumin, antibodies-online GmbH) for 5 min. Calcium bicarbonate solution, freshly prepared as described above, was used as a washing buffer. After three washes the samples were incubated with an 1:50 dilution of the secondary Anti-Rabbit IgG gold-labeled antibody (Sigma, Germany) for 5 min. In the last washing step, the washing buffer was gradually diluted with ultrapure water (1:0, 1:1, 0:1).

References

[1] E. Bäuerlein, *Biomineralization*, Wiley–VCH, Weinheim **2007**, ISBN 978-3-527-31641-0.

[2] S. Mann, *Biomineralization*, Oxford University Press, Oxford **2001**, ISBN 0-19-850882-4.

[3] H. A. Löwenstam and S. Weiner, *On Biomineralization*, Oxford University Press, New York **1989**.

[4] B. S. C. Leadbeater and R. Riding (eds.), *Calcification in the coccolithophorids Emiliana huxleyi and Pleurochrysis carterae, Biomineralization in lower plants and animals*, vol. 30, Systematics Association, Oxford University Press, Oxford **1986**.

[5] Y. Oaki and H. Imai, *Small* **2005**, *2* (1), 66.

[6] J. B. Thompson, G. T. Paloczi, J. H. Kindt, M. Michenfelder, B. L. Smith, G. Stucky, D. E. Morse and P. K. Hansma, *Biophys J* **2000**, *79*, 3307.

References

[7] J. L. Arias, D. J. Fink, S. Q. Xiao, A. H. Heuer and A. I. Caplan, *Int Rev Cytol* **1993**, *145*, 217.

[8] R. M. Sharp and H. Silyn-Roberts, *Biophys J* **1984**, *46*, 175.

[9] A. N. J. Heyn, *J Apl Phys* **1962**, *33* (8), 2658.

[10] Y. Nys and J. Dominguez-Vera, *Poultry Science* **2000**, *79* (1-2), 901.

[11] J. Gautron, M. T. Hincke and Y. Nys, *Connect Tissue Res* **1997**, *36*, 195.

[12] A. Awade, *Z Lebensm Unters Forsch* **1996**, *202*, 1.

[13] E. Holen and S. Elsayed, *Int Arch Allergy Immunol* **1990**, *91*, 136.

[14] J. Kelly, S. Locke, L. Ramaley and P. Thibault, *J Chromatogr A* **1996**, *720*, 409.

[15] A. Awade, S. Moreau, D. Molle, G. Brule and J. Maubois, *J Chromatogr A* **1994**, *677*, 279.

[16] J. Gautron, M. Hincke and Y. Nys, *Connect Tissue Res* **1997**, *36*, 195.

[17] S. Raz, P. C. Hamilton, F. H. Wilt, S. Weiner and L. Addadi, *Adv Funct Mater* **2003**, *13*, 480.

[18] S. Raz, S. Weiner and L. Addadi, *Connect Tissue Res* **2000**, *12*, 38.

[19] T. Y. J. Han and J. Aizenberg, *Chem Mater* **2008**, *20* (3), 1064.

[20] I. M. Weiss, N. Tuross, L. Addadi and S. Weiner, *J Exp Zool* **2002**, *293*, 478.

[21] E. Beniash, L. Addadi and S. Weiner, *J Struct Biol* **1999**, *125* (1), 50.

[22] L. B. Gower and D. J. Odom, *J Cryst Growth* **2000**, *210* (4), 719.

[23] M. J. Olszta, S. Gajjeraman, M. Kaufman and L. B. Gower, *Chem Mater* **2004**, *16*, 2355.

[24] Y.-Y. Kim, E. P. Douglas and L. B. Gower, *Langmuir* **2007**, *23*, 4862.

[25] S. J. Homeijer, M. J. Olszta, R. A. Barrett and L. B. Gower, *J Cryst Growth* **2008**, *310* (11), 2938.

[26] N. Loges, K. Graf, L. Nasdala and W. Tremel, *Langmuir* **2006**, *22*, 3073.

[27] Y. Kitano, *Bull Chem Soc Japan* **1962**, *35* (12), 1973.

[28] S. E. Wolf, J. Leiterer, F. Emmerling and W. Tremel, *J Amer Chem Soc* **2008**, *130*, 12342.

[29] M. Avrami, *J Chem Phys* **1941**, *9* (2), 177.

[30] M. Avrami, *J Chem Phys* **1940**, *8* (2), 212.

[31] M. Avrami, *J Chem Phys* **1939**, *7* (12), 1103.

[32] S. Chodankar and V. K. Aswal, *Phys Rev E* **2005**, *72* (4, 041931).

[33] S. Chodankar, V. K. Aswal, J. Kohlbrecher, P. A. Hassan and A. G. Wagh, *Physica B Cond Matter* **2007**, *298* (1), 164.

[34] H. M. Rietveld, *J Appl Cryst* **1969**, *2*, 65.

[35] V. Pipich, M. Balz, S. E. Wolf, W. Tremel and D. Schwahn, *J Am Chem Soc* **2008**, *130*, 6879.

[36] J. Rieger, T. Frechen, G. C. W. Heckmann, C. Schmidt and J. Thieme, *Faraday Discussions 136 on Nucleation and Crystal Growth*, Roy. Soc. Chem., London, 265–278.

[37] P. Somasundaran (ed.), *Encyclopedia of Surface and Colloid Science*, CRC Press, 2nd ed. **2006**, ISBN 978-0-8493-9615-1.

[38] M. Faatz, F. Gröhn and G. Wegner, *Adv Mater* **2004**, *16* (12), 996.

[39] L. Dai, X. Cheng and L. B. Gower, *Chem Mater* **2008**, *20* (22), 6917.

[40] O. Paris, C. H. Li, S. Siegel, G. Weseloh, F. Emmerling, H. Riesemeier, A. Erko and P. Fratzl, *J Appl Cryst* **2007**, *40*, 466.

[41] A. P. Hammersley, S. O. Svensson, M. Hanfland, A. N. Fitch and D. Hausermann, *High Press Res* **1996**, *14* (4-6), 235.

[42] R. Harris (ed.), *Electron microscopy in biology : a practical approach*, The practical approach series, IRL Press, Oxford Univ. Press, Oxford **1991**.

References

6 Cryogenic Molding of Nanotubes in Mesocrystalline Ice

Abstract Molding of nanotubes was accomplished by a bio-inspired cryo-molding approach, which avails the self-similar growth of hexagonal ice crystals. The branching during growth of hexagonal ice superimposed with the radial alignment of the ice crystals yields in a new variant of non-euclidian mesocrystals.

Reducing the mineral to a minority component, it is molden by the shock-freezing solvent. In this shock-frosting step, the solute is precipitated and occluded in the solvent's grain boundaries and is constrained to adopt a sheet-like shape. The subsequent freeze-drying removes the solvent and excavates the freshly formed sheets, which immediately roll up and form nanotubes featuring a remarkable aspect-ratio. The presented approach is not restricted to a specified chemical reactivity of the desired material, and is capable to mold inorganic aqueous solutions and suspensions. This morphogenesis does not involve chemical reaction and avoids complications and drawbacks like hazardous by-products or contamination.

Introduction

Molding on the nano- and micrometer scale can be regarded as one of the supreme disciplines of material science, as nano- and microstructured materials bear a wide range of application and may exhibit superior properties. Nature teaches lessons on this topic by showing an ultimate degree of sophistication and miniaturization in its biogenic materials. A prominent example is the nacreous layer of the gastropoda *Haliotis laevigata*, which consists of tectonic aragonite crystals embedded in an organic sheet-like matrix. The layer-directed growth of the aragonite tablets is only restricted by neighboring tablets, which are separated by occluded organic matrix at the end of mineralization. The timing during this process leads to a so-called "stack-of-coins"; in the case of bivalve shells, it yields in a "brick wall"-like structure.[1] This mesoscaled and layered structure with crystallographic oriented crystals features a toughness which is hundreds of times

6. Cryogenic Molding of Nanotubes in Mesocrystalline Ice

higher than pure aragonite. Its fracture behavior is dominated by the incorporated biopolymers like in the sea urchin spine (already discussed, *cf.* p. 2). An analogous structure, but on a much lower length scale, can be found in the case of the so-called 'transition bars', which form during the transformation of the amorphous PILP phase to the crystalline state. Likewise, the employed polymer is occluded in this intersticial transition bars.[2,3] All-day examples of tracery on a frosted window or snow-flakes are more familiar representatives of mesostructure and mesocrystallinity. The structure is dendritic, self-similar and is dominated by the growth and side-branching along the a_i-axes. These structures can be regarded as mesocrystals as they fulfill the definition: they represent mesoscopically structured crystals, whose subcrystals are aligned "in a common crystallographic register".[4] In short, ice crystal are employed in this chapter as a model system in order to explore the principle above of occlusion and the potential of the concomitant molding of impurities in mesocrystals.

A droplet of 4 µL water containing 'impurities' like salts or particles was shock-frozen in liquid propane, cryo-fractured and freeze-dried. The latter step of this morphogenetic process was followed by *in situ* cryo-scanning electron microscopy (cryo-SEM) where the obtained structures were characterized by transmission electron microscopy (TEM). Three exemplary representatives of 'impurities' were choosen: calcium carbonate as a classical mineral, cadmium sulfide as a classical II/IV-semiconductor, and a suspension of the simple insoluble ferrocyanide $K_2Zn_3[Fe(CN)_6]_2$ as a simplistic model of metal-organic frameworks (MOF, the ferrocyanide is referred to as sMOF hereinafter).

Results   Shock-freezing of a droplet in liquid propane generates a high, radially orientated temperature gradient. This induces the nucleation of hexagonal ice on the droplet surface, and the further growth along the a_i direction is orientated radially along the highest temperature gradient. One frozen droplet of a saturated calcium bicarbonate solution is shown in Fig. 6.1a; it already shows a radial orientation of the ice subcrystals which emerges more clearly after freeze-drying removed the subcrystals (Fig. 6.1b). The highly porous and ramified residue is composed of calcium carbonate. Some of the branches appear thicker; they form by a partial collapse of the ramified structure.

After a short phase of freeze-drying, a freshly cleaved ice surface shows already the intersticial occluded impurities which emerge initially as small parallel flashes (*cf.* Fig. 6.2 in case of cadmium sulphide). Further freeze-drying results in the exposing of sheets (Fig. 6.2b, here sMOF), which start to coil up immediately (Fig. 6.2c, here sMOF). Two examples of an intermediate state of this reeling process, which later yields in the formation of

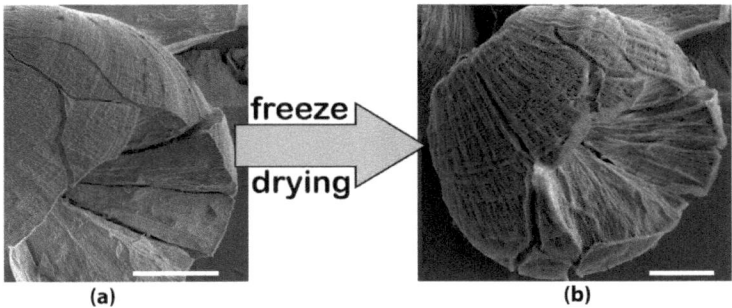

Figure 6.1 A droplet of saturated Ca(HCO$_3$)$_2$ solution **(a)** short after frosting, breaking, and **(b)** after accomplished freeze drying (20 min). Scale bars: 500 µm.

nanotubes, could be imaged by TEM (Fig. 6.3a and b, here sMOF). The reeling is induced by a different drying speed of the sheet's sides. If one side dries a bit faster, the whole sheet will bend as the faster drying side contracts with respect to the flip side.

The final state of this reeling process is the formation of nanotubes, which is shown in Fig. 6.4 in the case of calcium carbonate and sMOF. The sharp scattering contrast of the tubes' wall evidences their tubular character. These nanotubes can feature an exceptional aspect-ratio. They show a typical diameter of 75 – 100 nm but with an axial length of up to approximately 6 µm at the the same time (*e. g.* Fig. 6.4c, sMOF). Electron diffraction shows only weak Debye-Scherrer-rings (see insets in Fig. 6.4), which may imply that the tubes are amorphous but start to crystallize under radiative strain. Those findings could be ascertained by wide-angle X-ray scattering performed at a microfocus beamline equipped with an ultrasonic levitator. The levitation technique allowed a measurement of the X-ray diffraction pattern of a rotating but frozen droplet. In case of frozen calcium bicarbonate solution, no diffraction peaks were detected except the diffraction of hexagonal ice which vanish completely if the droplet is molten. The diffraction pattern of the frozen calcium bicarbonate solution corresponds with that of pure and shock-frozen water. (*cf.* Fig. 6.5)

To sum it up, a facile cryo-molding approach employing the mesocrystallinity of ice led to the formation of hollow nanotubes. This method is capable to mold inorganic aqueous solutions and suspensions. In the shock-frosting step (Fig. 6.6a – 6.6e), the solute is precipitated and occluded in the solvent's grain boundaries and is forced to adopt a sheet-like shape. Many hexagonal ice crystals nucleate on the droplet surface (Fig. 6.6b) and

6. Cryogenic Molding of Nanotubes in Mesocrystalline Ice

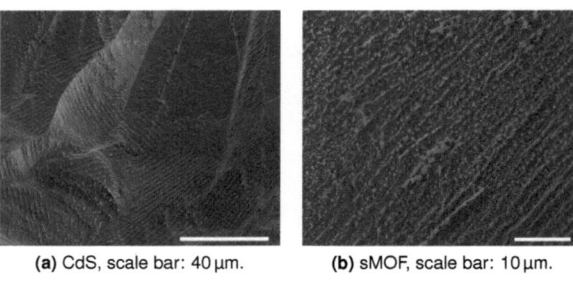

(a) CdS, scale bar: 40 μm.　　(b) sMOF, scale bar: 10 μm.

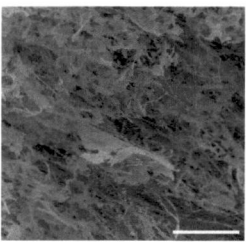

(c) sMOF, scale bar: 10 μm.

Figure 6.2 Scanning electron micrographs of intersticial deposited solute and suspensate showing grain boundaries of a frozen droplet.

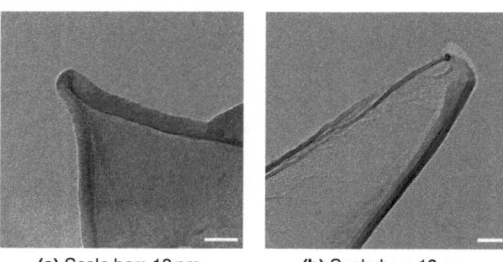

(a) Scale bar: 10 nm.　　(b) Scale bar: 10 nm.

Figure 6.3 Transmission electron micrographs showing the incipient reeling of a sMOF sheet, which later yields in the formation of a nanotube.

Summary and Conclusions

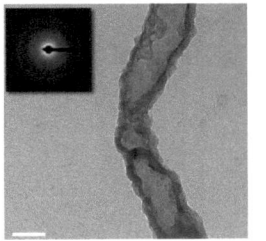
(a) CaCO₃; scale bar: 100 nm.

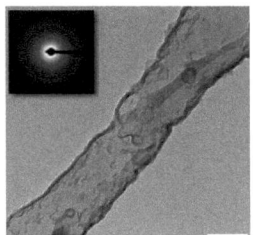
(b) sMOF, scale bar: 200 nm.

(c) sMOF, scale bar: 500 nm, inset: 250 nm.

Figure 6.4 Transmission electron micrographs of finally formed nanotubes.

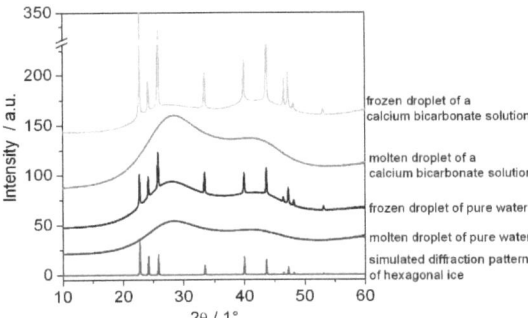

Figure 6.5 Wide-angle X-ray scattering of frozen and molten droplets containing saturated calcium bicarbonate solution or pure water.

61

6. Cryogenic Molding of Nanotubes in Mesocrystalline Ice

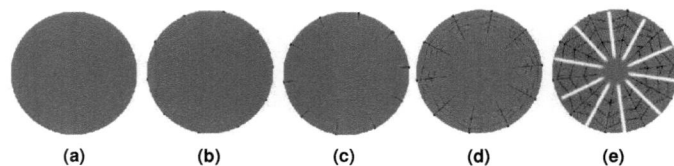

Figure 6.6 Scheme of the presented cryo-molding approach.
A droplet containing solute **(a)** is shock-frozen in liquid propane, which **(b)** induces the nucleation of hexagonal ice on the droplet's surface. **(c)** Due to the radial alignment of the temperature gradient, pure hexagonal ice crystals grow towards the center of the droplet. **(d)** Similar to snowflakes, branching occurs along the a_i axes. Ice crystallizes as a pure phase and the solute concentration in the yet non-solidified solution increases concomitantly. **(e)** In a late state, a critical solute concentration is reached and the solute precipitates. It is occluded in grain boundaries and therefore adopts a sheet-like shape which later show reeling if the ice is removed by freeze-drying.

start to grow rapidly orientating their a_i-axes radially (Fig. 6.6c). The solute concentration in the yet non-solidified solution increases concomitantly. Side-branching occurs, in analogy to the growth behavior of snow-flakes (Fig. 6.6d). At a critical solute concentration, the solute precipitates and is occluded in grain boundaries and adopts by this a sheet-like shape (Fig. 6.6e). The subsequent freeze-drying removes the frozen solvent and excavates the freshly formed sheets, which immediately roll up and form nanotubes featuring a remarkable aspect-ratio.

Strictly spoken, the symmetry of the microcrystalline ice is non-euclidian as it does not show a translational invariant, periodic three-dimensional order in the euclidian sense. The gradient of joint primary fields, which induced the crystallization of the hexagonal ice, is radially orientated towards the center of the droplet. The evolving structures are radially orientated as well. Furthermore, this radial alignment of crystal is superimposed by the dendritic and self-similar growth of the ice crystals by branching. The shock-frozen droplets represent a special case of mesocrystals, which were proposed to be named as 'bent or splayed mesocrystals' as they do not follow Euclidian geometry.[4]

Cryo-molding evolved recently as a promising technique providing access to a diverse array of complex structures; this morphogenesis does not require chemical reaction thus avoiding complications and drawbacks like hazardous by-products or contamination. Deville et al. exemplarily mimicked the structure of the nacrous layer employing a simple water-freezing method.[5] One-dimensional nanostructures have attracted considerable attention as they may serve as nanowires, nanoscaffolds in nanoscale electronics, optoelectronics or even tissue engineering.[6–9] Beside the well known synthetical approaches such as electrodeposition, laser ablation, templating or metal-organic chemical vapor deposition,[10–13] cryo-molding was shown to be applicable for molding of silica, organic polymers or nanoparticle suspensions to yield nanowires and porous fibers recently.[14–16] All the findings presented above clearly show that cryo-molding in frozen droplets is a exceptional facile way of nanotube formation. In addition, it is not restricted to special chemical reactivities of the desired material.

Experimental Part

Cadmium bisulfide and calcium bicarbonate solutions were prepared analogously as described in Chapter 3 on p. 26. The ferrocyanide suspension was prepared by mixing equimolar amounts of $K_4[Fe(CN)_6]$ and $ZnCl_2$ (0.05 mol l^{-1}) shortly before shock-frosting. The educt solution was shock-frosted by plunging the respective droplet in liquid propane, and transferring it immediately into a cryo-preparation chamber (FEI Quorum PolarPrep 2000). After propane was removed ($-60°$ C, 10^{-6} mbar), it was cryo-fractured. Afterwards the samples were transferred to the cryo-SEM (FEI xT Nova 600 Nanolab). Water was slowly removed by sublimation ($-5°$ C, 10^{-6} mbar) by means of a temperature-controlling cryo-stage. The ice sublimation and the concomitant exposure of cryogenic structures was monitored *in situ* (2–10 kV).

Transmission electron microscopy investigations were carried with a Phillips EM 420 running at 120 kV, equipped with an ORCA-ER camera (1024 × 1024 pixel) and run with AMT Image Capture Engine v5.42.540a. The samples were transferred on lacey-coated TEM grids (Plano, Germany) as obtained by freeze-drying followed by cryo-SEM. Wide-angle X-ray scattering experiments were performed using the µSpot beamline at BESSY, which is equipped with a double crystal monochromator, yielding a monochromatic beam ($\lambda = 1.00257$ Å, as calibrated with an external corundum standard).[17] A pinhole system provides a beam of 20 µm cross section with a photon flux of $\sim 10^9$ s^{-1} at a ring current of 100 mA. The scattering vector $q = 4\pi/\lambda\ \sin(\theta)$ is defined in terms of the angle 2θ between incident and scattered radiation of the wavelength λ. The obtained scattering data were processed by employing algorithms of the computer program FIT2D.[18]

References

[1] S. Mann, *Biomineralization*, Oxford University Press, Oxford **2001**, ISBN 0-19-850882-4.

References

[2] L. B. Gower and D. J. Odom, *J Cryst Growth* **2000**, *210* (4), 719.

[3] L. Dai, X. Cheng and L. B. Gower, *Chem Mater* **2008**, *20* (22), 6917.

[4] H. Cölfen and M. Antonietti, *Mesocrystals and Nonclassical Crystallization*, John Wiley & Sons Ltd, The Atrium, Southern Gate, Chichester **2008**, ISBN 987-0-470-02981-7.

[5] S. Deville, E. Saiz, R. K. Nalla and A. P. Tomsia, *Science* **2006**, *311*, 515.

[6] C. Y. Xu, R. Inai, M. Kotaki and S. Ramakrishna, *Biomaterials* **2004**, *25*, 877.

[7] X. Duan, Y. Huang, Y. Cui, J. Wang and C. Lieber, *Nature* **2001**, *409*, 66.

[8] M. R. Ghadiri, J. R. Granja, R. A. Milligan, D. E. McRee and N. Khazanovich, *Nature* **1993**, *366*, 324.

[9] S. Zhang, D. M. Marini, W. Hwang and S. Santoso, *Curr Opin Chem Biol* **2002**, *6*, 865.

[10] A. M. Morales and C. M. Lieber, *Science* **2002**, *279*, 208.

[11] L. Zhi, T. Gorelik, J. Wu, U. Kolb and K. Müllen, *J Am Chem Soc* **2005**, *127*, 12792.

[12] J. Yuan, Y. Xu, A. Walther, S. Bolisetty, M. Schumacher, H. Schmalz, M. Ballauff and A. H. E. Müller, *Nat Mater* **2008**, *7*, 718.

[13] M. P. Zach, K. H. Ng and R. M. Penner, *Science* **2000**, *290*, 2120.

[14] H. Zhang, J.-Y. Lee, A. Ahmed, I. Hussain and A. I. Cooper, *Angew Chem Int Ed* **2008**, *47*, 4573.

[15] W. Mahler and M. F. Bechtold, *Nature* **1980**, *285*, 27.

[16] J. T. McCann, M. Marquez and Y. Xia, *J Am Chem Soc* **2006**, *128*, 1436.

[17] O. Paris, C. H. Li, S. Siegel, G. Weseloh, F. Emmerling, H. Riesemeier, A. Erko and P. Fratzl, *J Appl Cryst* **2007**, *40*, 466.

[18] A. P. Hammersley, S. O. Svensson, M. Hanfland, A. N. Fitch and D. Hausermann, *High Press Res* **1996**, *14* (4-6), 235.

7 Nonclassical and Symmetry-Breaking Phase Selection of Calcium Carbonate Triggered by Amino Acids

Abstract A nonclassical phase selection of calcium carbonate polymorphs is triggered by the handedness of amino acids if they are used as additives during calcium carbonate crystallization. In case of the natural levorotatory enantiomer, the less stable polymorphs aragonite and vaterite were formed. In presence of the dextrorotatory enantiomer, the stable polymorph calcite is formed and an Ostwald ripening was identified. The arrangement of amino acids absorbed at $(104) \times (10\bar{4})$ growth steps were modeled molecularly. Growth steps from the {104} set of crystal planes show stereoselectivity which is maybe caused by the exposure of chiral crystal (214) crystal planes at {104} growth steps. Cryptochiral contaminants, which were suspected to be responsible for the seemingly symmetry breaking, are shown to suppress the effect of phase selection rather than promoting it. The situation prior to nucleation is populated by large numbers of coexisting enantiomeric and diastereomeric amino acid complexes. It is discussed whether the weak parity violation energy difference (PVED) may cause the symmetry breaking. However, no experimentally evidence could be achieved because of insufficient significance of the available and applied methods. Neither by means of circular dichroitic deviations nor the disappearance of phase selection in deuterated media was observed in statistical significance. This phase selection represents the first example of symmetry breaking phase selections and can be queued up with other examples of symmetry breaking and thus illustrates a totally different kind of nonclassical crystallization of calcium carbonate.

Associated *Angew Chemie Int Ed* **2007**, *46* (29), pp. 5618–5623,
publications resp. *Angew Chemie* **2007**, *119* (29), pp. 5716–5721.

Angew Chemie Int Ed **2008**, *47*, (20), pp.3683–3686,
resp. *Angew Chemie* **2008**, *120*, (20), pp. 3741–3744.

Nonclassical Symmetry-Breaking Phase Selection of $CaCO_3$ by Amino Acids

Introduction Chirality and the emergence of homochirality are among the most intriguing and inspiring phenomena in nature.[1] There are many attempted explanations,[2,3] where one of them focuses on the chiroselective adsorption of amino acids onto chiral mineral surfaces, in particular on the common rock-forming mineral calcite.[4] Since the crystal surface lacks the symmetry features of the bulk crystal, the adsorption of an achiral molecule onto a crystal surface may produce chiral arrangements in two dimensions.[5–13] *Vice versa*, a chiral molecule on a crystal surface may lead to a chiral entity when ordered two-dimensional adlayers are formed.[14–19]

Calcite, which is the thermodynamically stable of the six known calcium carbonate polymorphs, was presumably the most abundant marine mineral in the Archaean era, about 3.8 to 25 billion years ago. Sumner stated that calcite "precipitated as crystals directly on the sea floor".[4] The selective binding of different amino acids onto crystal surfaces has been reported for calcium carbonate,[20] but also for copper[14–17] or other examples.[18,19] The chiroselective absorption of amino acids onto calcite was demonstrated by Hazen and coworkers,[13] and the formation of chiral morphologies through the selective binding of D- and L-aspartate on growth steps of calcite was studied by Orme et al.[20] These studies provided convincing evidence that the shape of calcite crystals can be modified by the binding of chiral molecules through stereochemical recognition.

The shape of centrosymmetric crystals, like the trigonal calcite (space group $R\bar{3}c$) as well as the orthogonal aragonite ($Pmcn$) and vaterite* are delimited by pairs of faces $\{hkl\}$ and $\{\overline{hkl}\}$, which are linked in case of chiral $\{hkl\}$ faces by inverted chirality of the $\{\overline{hkl}\}$ face. The rhombohedral morphology of calcite is delimited by a $\{104\}$ set of crystal planes, which possess under ideal conditions mirror symmetry, like all $\{h0l\}$ face must hold in the $R\bar{3}c$ space group. The (104) planes themselves show pg symmetry.[25] Any rhombohedral crystal shape possesses mirror symmetry along the rhombus diagonal, parallel to the c-axis, if it belongs to a a trigonal space group.[26] Although, the likewise common scalenohedral calcite shape which are delimited by $\{214\}$ faces are in fact chiral. In reality, a non-ideal (104) face, will show in local detail chirality due to the presence of steps and kink sites.[†, 26] Taking a closer look on the growth steps, the local symmetry characteristics of a single step evolve from the combination of surfaces and can partially expose other chiral crystal planes like the ones of the $\{214\}$ set.

---

* The exact structure of vaterite is still hardly debated; for instance, its space group is still unclear: $Pnma$, $P6_3/mmc$ and 6_322 are proposed to date.[21–24]

† Everything is triclinic if you look at it hard enough. (Charlie Burnham)

The paradigm of stereochemical recognition, which was introduced and developed two decades ago,[27,28] is a central tenet in the field of biomineralization.[29] It states that specific crystal surfaces are stabilized through the binding of molecules such as peptides and proteins due to the stereochemical match of the adlayer and the crystal lattice lowers their surface energies.[30–36] This template model found ample support in a series of investigations of the macroscopic shapes of organic or calcium carbonate crystals.[37–44] Still, crystal shape is not only depending on surface energy but also on growth kinetics. During the past decade a number of elegant studies have related crystal shape to the growth kinetics which is governed by coordination at kinks and on atomic ledges rather than on the flat faces.[20,45–47] In this chapter, the specific chiral interactions in the course of mineralization evolve to be a dominant factor and can exert an important influence on the phase selection of calcium carbonate.

Crystallizations of calcium carbonate were performed with enantiopure amino acid as additives applying the slow diffusion technique. In a representative experiment, either enantiomorph of the corresponding amino acids was dissolved in aqueous solutions of 10 mmol l^{-1} of calcium chloride. An adjustment of the *p*H was abstained in order to eliminate possible effect of foreign ions on the precipitation of calcium carbonate. The solution were incubated simultaneously together with 14 g of freshly ground ammonium carbonate in a desiccator. The crystallization was carried out at room temperature and was stopped after 48 h. The homogenous precipitates were collected, carefully washed, dried and further investigated.

The crystallization experiments were performed with all natural amino acids along with several synthetic ones. The achiral amino acid glycine served as a reference. Scanning electron microscopy (SEM) images of samples precipitated with the D- and L-enantiomers of alanine, proline, α-amino butyric acid, and aspartic acid as representatives of simple chiral amino acids are displayed in Fig. 7.1. Already alanine, the most simple chiral amino acid, shows a pronounced effect. In presence of the natural levorotatory form, the dominating phases are aragonite and vaterite. In the case of the dextrorotatory form, only calcite was obtained (Fig. 7.1a and b). Addition of D- or L-aspartic acid yielded identical results. Higher nonfunctional amino acids lead to the formation of aragonite resp. calcite: L- and D-proline lead to the crystallization of aragonite (Fig. 7.1c and d). Likewise, L- and D-α-amino butyric acid (Fig. 7.1e and f) select calcium carbonate phases in the same manner. Amino acid bearing higher functionality like lysine or glutamic acid deviate from this phase selection behavior and amino acids with higher steric demands like tryptophan, tyrosine or *tert*-leucine do not exhibit phase selectivity at all, which may be attributed to a steric hinderance during the phase selection process. In comparison, crystallizations in presence of glycine or racemic alanine under otherwise identical experimental conditions yielded mixtures of calcite and aragonite (Fig. 7.2). Both are known to stabilize vaterite,[48,49] however the conditions chosen in the present case led to the formation of polymorph mixtures.

The phase selection was monitored at a quantitative scale by time-resolved measurement of pH and calcium concentration ($[Ca^{2+}]$) in the supernatant and by X-ray diffraction (XRD). The pH profile of the solution during the crystallization in the presence of D- or L-alanine is depicted in Fig. 7.3. After starting of the reaction, the pH of the solution rises within an induction period of about 4 h from slightly below 7 to 8.9 due to the dissolution of ammonia, which is formed during the decomposition of $(NH_4)_2CO_3$. The better solubility of ammonia compared to that of CO_2 leads to the observed change in pH. This equilibrium adjusts within the remaining duration of the experiment until the NH_3 vapor pressure in the gas phase matches the ammonia partial pressure of the solution. The continuous formation of $CaCO_3$ is indicated by the plateau of the pH value between 1.5 h and 3 h, which is caused by the equilibrium of CO_2 uptake and HCO_3^- depletion of the solution due to the incipient precipitation of calcium carbonate. This depletion gives rise to a release of protons and balances the pH rise from the uptake of ammonia. For D-alanine the precipitation occurs earlier. The plateau is reached after a shorter period of time and at a lower pH level. For L-alanine, calcium carbonate precipitates later, i.e. a higher degree of supersaturation is needed to induce nucleation. After the completion of the experiment, the calcium concentration $[Ca^{2+}]$ of the supernatant was determined by atomic absorption spectroscopy (AAS) and yielded 0.8 mg l^{-1} in case of D-alanine resp. 1.3 mg l^{-1} in case of L-alanine. The time-dependent changes of $[Ca^{2+}]$ were monitored by means of a $[Ca^{2+}]$-sensitive electrode and revealed a considerable distinction between both crystallizations: L-alanine shows a uniform crystallization profile of an one-step steady-going precipitation, which leads to the metastable vaterite polymorph. In case of presence of D-alanine, the precipitation is a two-step process; here a redissolution of already precipitated calcium carbonate and a subsequent anew precipitation is traced. This represents a perfect example of the Ostwald law of stages.[50]

Fig. 7.4 depicts the experimental and fitted powder XRD patterns corresponding to the material collected from the mineralization experiments using valine as an additive. From the Rietveld scale factors,[51] it is possible to obtain the relative amounts of the different phases. In case of the D-enantiomer, only calcite was observed, whereas aragonite was found for the L-enantiomer. From such a Rietveld analysis, it is found that the weight fractions of the $CaCO_3$ polymorphs vaterite and aragonite in the presence of L-alanine are $\approx 3:1$. The results for D- and L-proline, α-amino butyric

(a) L-alanine, scale bar: 50 µm.

(b) D-alanine, scale bar: 200 µm.

(c) L-proline, scale bar: 50 µm.

(d) D-proline, scale bar: 200 µm.

(e) L-α-amino butyric acid, scale bar: 200 µm

(f) D-α-amino butyric acid, scale bar: 200 µm

Figure 7.1 Scanning electron micrographs demonstrating the phase selection in presence of different amino acids.

Figure 7.2 Scanning electron micrographs of crystallizations by addition of **(a–c)** achiral amino acids and **(d–f)** sterically hindered amino acids.

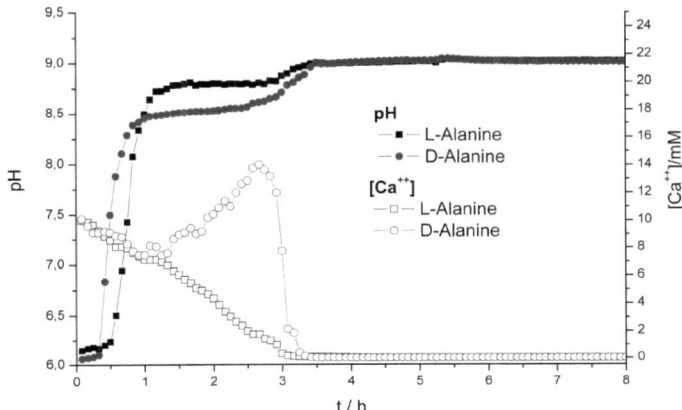

Figure 7.3 Progression of *p*H and [Ca^{2+}] during a crystallization with L- and D-alanine as additives.
The calcium concentration excesses the nominal calcium content (10 mmol l^{-1}) in the time-frame from 2 h to 3 h. This is due to the fact that the calcium-sensitive electrode is sensitive to H$^+$ and NH$_4^+$ as well, which are released resp. formed massively during the incipient redissolution (*vide infra*).

acid, aspartic acid and glycine are compiled together with those for D- and L-alanine in Table 7.1.

The growth of the calcite polymorph is known to be prevented by the strong surface binding of additives, which block the further transport of bulk material.[31] Bidentate acidic additives like amino acids attach more firmly to the growth steps than monofunctional additives like monofunctional carboxylic acids or amines do. Calcite crystals, which were grown in the presence of these ampholytes, develop characteristic edge defects and thus show the same behavior like simple monofunctional carboxylic acids or amines.[52] If the carboxylated fluorescent dye fluoresceine is used as an additive, an analysis of the obtained CaCO$_3$ crystals by means of confocal laser scanning microscopy (CFLSM) gives clear evidence for a surface functionalization of rhombohedral-shaped calcite crystals (Fig. 7.6). Red fluorescent crystals demonstrate the coverage of the crystal surface by the fluorophor and show as well characteristic edge defects. Still, unequivocal experimental proof for the formation of amino acid adlayers on the crystal surfaces is difficult provide as D-/L-amino acids carrying fluorescent groups do not induce a switch in phase because of the sterical hinderance which is associated with such functional groups (*vide supra*). Thermogravimetric analysis and

enantiomer	glycine*	racemic alanine		
achiral / racemic	100/0/0	84/0/16		
	alanine	α-amino butyric acid	proline	valine
L	27/0/73	0/100/0	19/81/0	0/100/0
D	100/0/0	100/0/0	100/0/0	100/0/0

* achiral amino acid

Table 7.1 Phase distribution (calcite/aragonite/vaterite) in weight percent induced by the presence of 1 mg ml^{-1} of the respective amino acids as determined by Rietveld refinement.

surface-sensitive X-ray photoelectron spectroscopy could not trace any organic material. However, ^{13}C-CP/MAS-NMR spectra of carefully washed bulk CaCO$_3$ crystals obtained in the presence of D- and L-alanine revealed weak signals of the amino acid. Elemental analysis of CaCO$_3$ precipitates obtained in the presence of alanine showed congruently a slight increase in nitrogen content (L: 0,05%, D: 0,08%). One may deduce from these findings that amino acids are bound to surface step edges and kinks but do not form a complete adlayer, which could be seconded by molecular modeling simulations (Fig. 7.7).

In order to probe the hypothesis that calcium cations on the crystal surface coordinated by bidentate amino acid ligands play a pivotal role in the observed phase selection, experiments were performed in which α-alanine was replaced by achiral β-alanine. In contrast to experiments using the achiral glycine (in which mainly calcite was formed), the addition of achiral β-alanine resulted in the preferential formation of aragonite needles with minor calcite contaminations (see Fig. 7.2). The switch from an α- to a β-amino acid changes the opening angle of the bidentate ligand and may improves the complex stability and thus the growth inhibition and thereby leads to a preferential formation of aragonite.

Molecular Modeling Ab initio forcefield calculations based on the COMPASS forcefield (Materials Studio)[53] approved this in the present case and, moreover, the increase of binding energy in case of amino acids and analogue ampholytes glycine (GLY), β-alanine (BALA), γ-amino butyric acid (GABA) and anthranilic acid (ANS) exceeds the sum of the binding energy of the monodentate carboxylic and amino additives (see Fig. 7.5).

Growth steps from the {104} set of crystal planes may uncover partially chiral crystal planes of the {214} set which would allow a stereoselective binding motif which could be shown by further *ab initio* forcefield calculations based on the COMPASS forcefield (Materials Studio).[53] Amino acids and their

Molecular Modeling

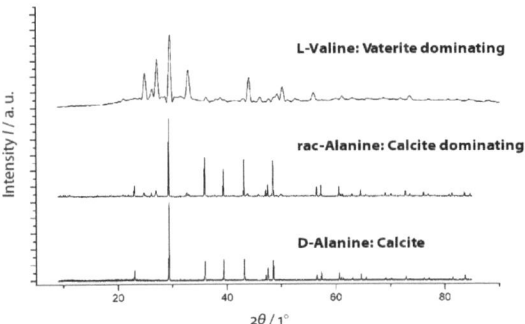

Figure 7.4 Diffraction patterns of calcium carbonate obtained under addition of 1 mg ml^{-1} L-, D- or racemic alanine.

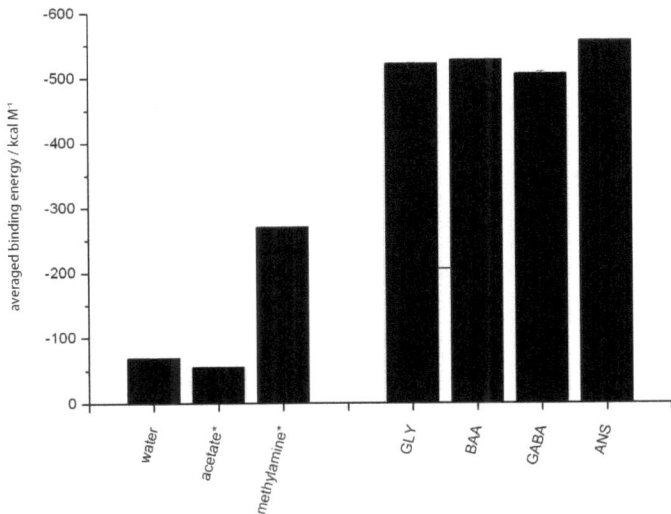

Figure 7.5 Comparison of binding energies of the monofunctional additives (methylamine and acetate) and the bidentate ampholytes glycine (GLY), β-alanine (BALA), γ-amino butyric acid (GABA) and anthranilic acid (ANS).
* Values for methylamine and acetate kindly provided by N. Loges.

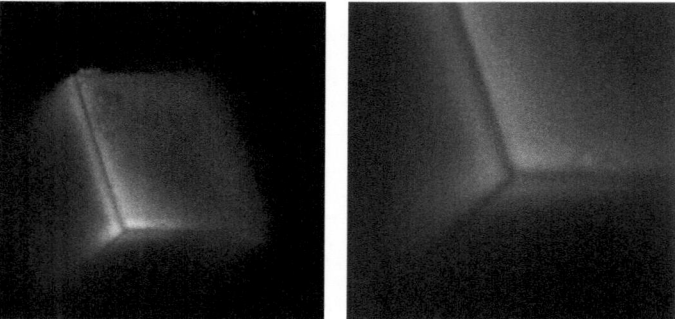

Figure 7.6 Confocal laser scanning microscopy images of calcium carbonate which were grown in the presence of 1 mg ml^{-1} fluoresceine.

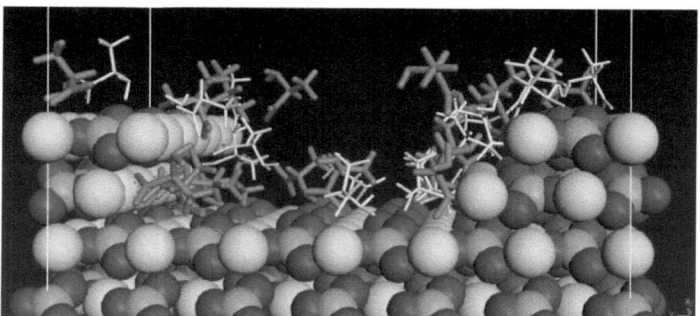

Figure 7.7 Binding motif of a racemic mixture of alanine, computed with the COMPASS force field. The L-enantiomer is colored in green and prefers coordination at the left step.

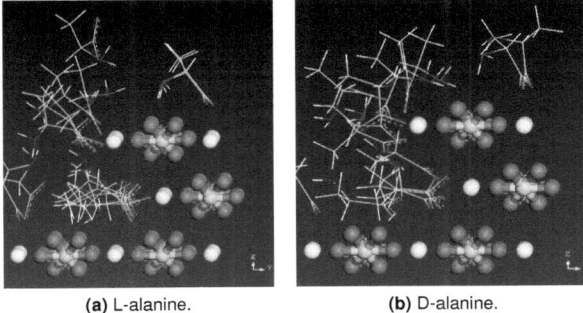

Figure 7.8 Detailed view of the calculated binding motif of **(a)** L-alanine and **(b)** D-alanine attached to (104) growth steps.

analogs were found to prefer a bidentate interaction by their amino and carboxylate groups and, as expected, the binding of the D- and L-enantiomers differs in the orientation of the residue at the asymmetric carbon with respect to the surface. In order to compare the binding strength for both enantiomers, the chemisorption of a racemic mixture of D- and L-alanine to a $(104) \times (01\bar{4})$ surface step was simulated and showed L-alanine to adsorb preferentially to the z-shaped growth step (Fig. 7.7). Furthermore, the L-enantiomer adopts much better the symmetry of this z-shaped surface step than its counterpart D-alanine (Fig. 7.8). Thus, the binding energy of one amino acid enantiomer is slightly higher, depending on the particular step.

The result of chiroselective accumulation at stepped growth features touches a possible mechanism for the prebiotic synthesis of homochiral amino acids and polypeptides and has implications for chiral catalysis as well.[10–13] The highly selective concentration of enantiomers along step-like features may favor an alignment of homochiral amino acids which is crucial to promote a homochiral polymerization which is considered to be a key step in the synthesis of self-replicating peptides.[54–56] Mineral-mediated chiral selectivity, in conjunction with homochiral polymerization, may thus provide a link between prebiotic synthesis and the RNA-protein world.

The Issue of Symmetry-Breaking However, the biggest question mark remaining is that of the chiro-dependent phase selection. Only if an unequal distribution of enantiomeric bindings sites are uncovered by growth steps or screw dislocations throughout the growing calcite phase, a chiro-dependent blockade of the growth would be possible. Then, one enantiomer would block the calcite growth more efficiently than the other and would thus lead the formation of less stable mineral phases. In terms of Ostwald's law of stages,[50] the activation barrier of the last phase transformation could be increased then by the addition of this enantiomer because a further attachment of growing material is blocked. But the assumption that an unequal distribution of binding sites throughout the growing material is present would break symmetry.[57–59] Taking into account that the three main polymorphs are centrosymmetric, the statistical distribution of different growth steps and the corresponding chiral binding sites should be equal, and the growing process should not be affected by a difference in handedness of the applied enantiomers. Now let us investigate the situation in a solution containing both calcium ions and amino acids. This situation, prior to phase selection and in fact even prior to nucleation, is difficult and unsettled due amino acids acting as ligands in the coordination chemistry of calcium.[60] The nucleation and crystal growth in the presence of amino acids (Haa) is dictated kinetically and thermodynamically by calcium amino acid complex formation equilibria such as the following.

$$[Ca(H_2O)_n]^{2+} + aa^- \rightleftharpoons [Ca(aa)(H_2O)_{n-m}]^+ + m\ H_2O$$
$$[Ca(aa)(H_2O)_n]^+ + aa^- \rightleftharpoons [Ca(aa)_2(H_2O)_{n-m}] + m\ H_2O$$
$$[Ca(aa)_2(H_2O)_n] + aa^- \rightleftharpoons [Ca(aa)_2(H_2O)_{n-m}]^- + m\ H_2O$$

Large numbers of enantiomeric and diastereomeric amino acid complexes may coexist and it seems plausible that five- or seven-coordinate species exist as well, although only sixfold-coordinated metal centers are depicted in Fig. 7.9. Most of the possible calcium complexes have neither been demonstrated in solution nor isolated and structurally been characterized, but they will exert a distinct influence on crystal growth and dissolution, as they have different structures, symmetries and thermodynamic stabilities. For instance, the most simple mono-alaninato calcium compound $[Ca(ala)(H_2O)_n]^{2+}$ is only present in small amounts for pH lower than 8.5, but if the pH values are higher than 11, then 16% of the total solved calcium will exist in this complex. Employing positive-ion ESI mass spectroscopy of alanine and calcium containing solutions show clearly the presence mono-, di- and tri-substituted calcium centered alaninato complexes. Prominent signals at higher masses indicate the existence of compounds with more than one calcium ion.[61] The suspicion of Lahav et al.,[25] that cryptochiral

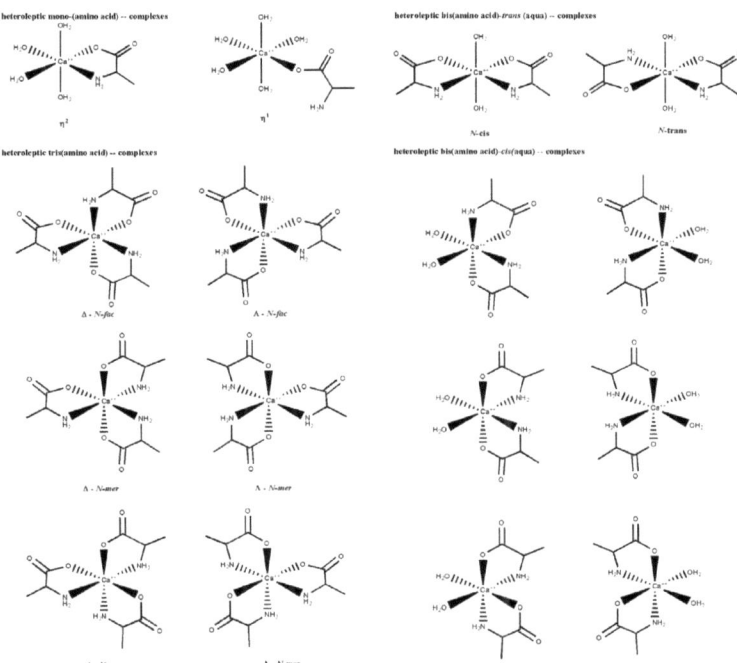

Figure 7.9 Selection of conceivable octahedral calcium-alaninato coordination compounds. The stereochemistry of the alanine ligand is not shown.

contaminations* trigger this phase selection, could be disproven by different approaches. In order to probe such influence of chiral contaminants, several crystallizations were carried out in presence of enantiopure amino acids with and without addition of traces of a second enantiopure amino acid of either handedness, which lead in any case to a complete loss of phase selectivity. Thus, homochiral impurities clearly suppress the effect of phase selection rather than promoting it.[61]

The existence of numerous variants of calcium complexes, which are all present in the mother solution linked together by a weak equilibrium, may hold the key to this phase selection, which apparently breaks symmetry.

The tri-substituted octahedral complexes split up into two enantiomers Δ and Λ, if the absolute λ- and δ-conformations of the chelate ring is neglected

---

* cryptochiral contaminants: analytically unverifiable impurities.[59]

(Fig. 7.9). In case of asymmetric ligands like the N,O-bis-chelate complex [Ca(aa)$_3$]$^-$ the *fac*- and *mer*- isomers have to be taken in account. Thus, the solution contains a large number of enantiomeric and diastereomeric calcium amino acid complexes besides the besides the minority component of non-coordinated amino acids. In order to generate different physical and chemical effects, the ratio of diastereomers Λ/Δ of these complexes must change if the handedness of the amino acid ligand is altered.

In the following discussion, we abbreviate complexes of different symmetry as follows:

$$\Lambda - [Ca(\text{L}-\text{aa})_3]^- := \Lambda\text{L}$$
$$\Delta - [Ca(\text{L}-\text{aa})_3]^- := \Delta\text{L}$$
$$\Lambda - [Ca(\text{D}-\text{aa})_3]^- := \Lambda\text{D}$$
$$\Delta - [Ca(\text{D}-\text{aa})_3]^- := \Delta\text{D}$$

Following symmetry, the enthalpy of complex formation ΔG_{form} of mirror related complexes have to be equal ($\Delta G_{form}(\Lambda\text{L}) = \Delta G_{form}(\Delta\text{D})$). Differences in complex stability will favor one of the diastereomers of the same amino acid enantiomer (e.g. $\Delta G_{form}(\Lambda\text{L}) < \Delta G_{form}(\Delta\text{L})$). In the case of the other amino acid enantiomer, the ratio should reverse (then $\Delta G_{form}(\Lambda\text{D}) > \Delta G_{form}(\Delta\text{D})$). Self-evidently, diastereomeric complexes like ΛL and ΔL cannot interact identically with a chiral binding site of calcite. However, no symmetry breaking should be present and the ratio of Λ/Δ-diastereomers in each solution should correspond and, naturally, the Λ/Δ ratio should inverses if the handedness of the amino acid is changed. Therefore, the both favored complexes are related by mirror symmetry.

However, articles concerning theory and observed symmetry breaking are nowadays legion. In case of the occurrence of chiral resolution during the crystallization of different chiral chlorates and bromates,[57-59] the cause of this symmetry breaking could be tied down to a puzzling interplay of nucleation, convection[57] and stirring[58] based on the principle of the common-ancestor effect[62] which leads to an advection-mediated chiral autocatalysis. Furthermore, the weak parity violation energy difference (PVED) arising from parity non-conserving neutral currents is now widely accepted.[63] The energy difference found for amino acids is $\sim 10^{-17} k_B T$.[64] The theoretical value of PVED is highly sensitive to the quality of the applied methods and may increase by up to two orders of magnitude.[65,66] In principle, it could be possible to amplify this small energy deviation under conditions like polymerization, condensation or in particular crystallization in order to affect chemical behavior and to prove the possible influence of the PVED on chemistry.[64]

The PVED is a function of the atomic number Z of the asymmetric atom to the power of six. Indeed, the crystallization behavior of sodium ($Z = 6$), cobalt ($Z = 30$) and iridium ($Z = 77$) centered complexes with respect to their Λ/Δ-conformation featured a breaking of symmetry during their crystallization.[67] Scolnik et al. based their approach on amplification the PVED effect by homochiral polymerization,[68] and they actually found subtle differences in circular dichroism (CD) and isothermal titration calorimetry (ITC) depending on the chirality of the monomers. This substantiates the assumption that an intramolecular autocatalytic amplification of a slight chiral deviation takes place. Interestingly, the effect vanishes under addition of 80% of deuterated water. Bulk water can be viewed as a mixture of *ortho*-water and *para*-water in ratio 3:1. *Ortho*-water represents the case, in which the proton spins are parallel and therefore bears a magnetic field. The authors propose that due to the magnetic component of *ortho*-water the L-enantiomer is preferentially solvated and that if mainly deuterated water is present the spin isomers are 'scrambled'.[68–70] Thus if the observed effect vanishes in highly deuterated water, this will indicate, whether PVED plays a role in the observed effect or not.

Calcium-centered complexes ($Z = 20$) can be considered to be amino acid oligomers and may exhibit comparable behavior. But in the present case, a superposition of enhancement effects could occur: *(a)* The enhancement of PVED due to the crystallization of chiral complexes like the results of Szabo-Nagy et al.[67] *(b)* The enhancement of PVED because of an oligomerization of chiral amino acids as in the case Scolnik et al. presented,[68]. *(c)* A subtle difference in solvation and protonation of amino acids as discussed by Scolnik et al. which may lead to a distinct deviations in complexation behavior and pH profile.[68] But a change in the pH profile can lead to the observed switch in phase. Several experiments utilizing other crystallization methods with differing pH profiles proved that this leads to distinct differences in the obtained $CaCO_3$ phase composition.

Following Scolnik et al.,[68] CD spectra of different amino acids were recorded under addition and absence of calcium but under otherwise complete identical conditions. However, no deviation between calcium-free and calcium-doped solutions could be detected within the precision limits of the instrument. The spectra of proline are shown in Fig. 7.10, accompanied by the spectra of L-alanine. The inset demonstrates the uncertainty of the conducted measurements. Additionally, crystallizations in 80% deuterated water were carried out in presence of the enantiomers of alanine, proline, and glutamic acid, but they did not differ from samples received from bulk water (Fig. 7.11). However, these crystallizations had to be carried out in very reduced volumes which lead to a severe reduction in phase selectivity for both bulk and deuterated water due to the distinct change in the pH evolution. However, no substantive experimental proof of the dependence of the Λ/Δ-diastereomeric ratio on the handedness of the amino acid ligand could be achieved applying available methods.

Summary Summing up the results, they demonstrate that chiral additives selectively can lead to a nonclassical crystallization in terms of a symmetry breaking phase selection. Nucleation and crystal growth are not only an expression of equilibrium energetics, but also of the growth kinetics. Only the levorotatory additives allowed growth kinetics to dominate the crystallization process, thereby leading to a nonclassical phase selection of a metastable $CaCO_3$ polymorph. It was demonstrated for the first time that the phase selection of $CaCO_3$ can depend on the chirality of the additives. This contribution explores chiral additives in polymorph control of minerals for the first time and addresses nonclassical polymorph selection due to symmetry breaking. Based on this study, one may speculate further wether the homochirality of amino acids is based on the difficult and puzzling interplay of chirality and complex formation. Then, calcium-centered amino acid complexes could have shifted the enantiomeric equilibrium towards the nowadays dominating L-enantiomer due to the PVED effect.

Experimental Part

Ultrapure water was used in each step ($>18.3\,\mathrm{M\Omega}$, Millipore Synergy 185, UV oxidation). Glass slides which were used for sample collecting were cleaned in a mixture of ammonia solution (28 – 30%), hydrogen peroxide in water (1:1:5 by volume) for 10 min at 80° C. Afterwards, the slides were rinsed with ultrapure water and dried with nitrogen (99.999%). Crystallization was carried out employing the so-called slow-diffusion technique.[71–77] An amount of 250 mg of each amino acids enantiomer (Acros or Sigma-Aldrich, >98%) was dissolved in 250 ml 10 mmol l^{-1} $CaCl_2$ solution (Merck, Suprapur), in some cases by means of ultrasonification. Both solutions had a typical pH of 7 as an adjustment of pH before starting the crystallization was abstained, because foreign ions can strongly affect the precipitation of calcium carbonate. The solutions were incubated simultaneously with 14 g of freshly ground ammonium carbonate (Acros, *p. a.*) in the same desiccator at room temperature for a period of 48 h. After completion, the slides were gathered, cleaned with water in order to remove weakly adhered crystals. Crystallizations in deuterated water were performed in small perforated snap cap vials (10 ml volume). A solution of 1 mg ml^{-1} amino acid was mixed with 0.1 ml 500 mM $CaCl_2$ and 4 ml of H_2O resp. D_2O (Deutero, 99.9%). Then, six samples were incubated

Experimental Part

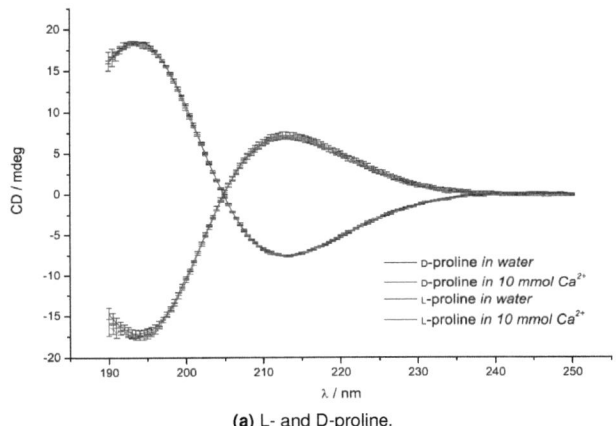

(a) L- and D-proline.

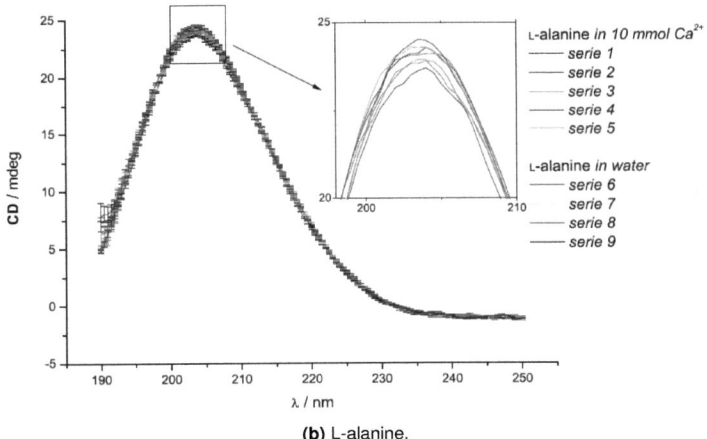

(b) L-alanine.

Figure 7.10 CD spectra of proline and alanine under addition and absence of 10 mmol l^{-1} calcium chloride. The inset demonstrates the deviation of different experimental series, which does not show uniform behavior with regard to the presence of calcium.

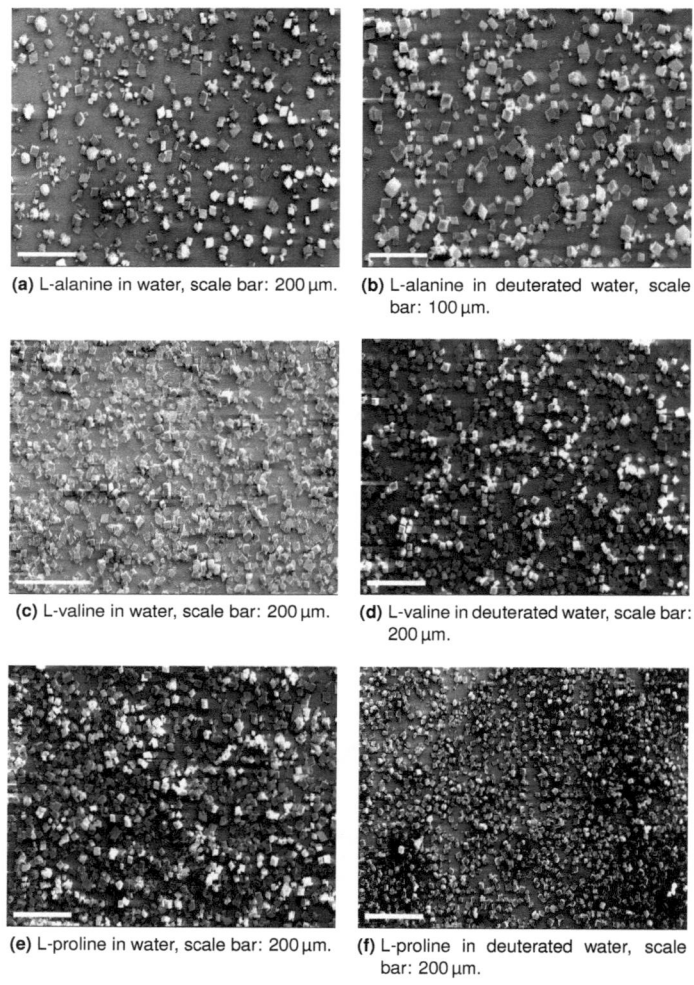

Figure 7.11 Scanning electron micrographs of crystallizations under addition of 1 mg ml^{-1} L-amino acids either in water or in 80% deuterated water.

together in a desiccator for 48 h on accurate equal distance concerning the Petri dish with 5 g of freshly ground ammonium carbonate. Furthermore, several crystallizations were conducted with methods of Cölfen and Kitano which are characterized by different pH profiles.[78,79]

The calcium concentration was determined by means of atomic absorption spectroscopy in the supernatant solution of crystallization experiments employing alanine enantiomers as additives. In case of the levorotatory form, the supernatant contained 1.3 mg ml^{-1} Ca^{2+} whereas in case of the dextrorotatory enantiomer it contained 0.8 mg ml^{-1}. A WTW SenTix 81 pH electrode with automatic temperature compensation, a [Ca^{2+}]-sensitive WTW Ca800-electrode, and two WTW pH/Ion 340i processing units were used to monitor the pH and [Ca^{2+}] in intervals of 5 min. Elemental analysis of the precipitates yielded in a blank test 0.05% N, in presence of L-alanine 0.06% N, in presence of D-alanine 0.06% N. Samples of CaCO$_3$ obtained in presence with L- and D-alanine were measured by means of CP/MAS-NMR with acquiring data time of 4 d. X-ray diffraction patterns were recorded in transmission mode with CuKα_1 radiation using a Siemens D5000 equipped with a Braun M50 PSD. Scanning Electron Microscopy (SEM) was performed with a Zeiss DSM 940 (acceleration voltage 3–15 kV, working distance 5–7 mm). For better conductivity, the samples were sputtered with 10 nm of gold by means of a Baltec MED020 coating system.

For molecular modeling and dynamic simulation, Materials Studio v4 from Accelrys was employed.[53] An initial comparison between the results of *ab initio* calculations with DFT (PB91, GGA) and the COMPASS forcefield confirmed that electrostatic interactions dominate and that the computational results are well reproduced using the less time consuming COMPASS forcefield. The additive geometry of L- and D-alanine were energetically optimized with Forcite, COMPASS forcefield and an atom-based summation method. Start-up geometries of the bulk calcite crystal were obtained by cleaving a (104) surface at a depth of 21.25 Å from a 2 × 7 supercell. The bulk crystal was constrained to fixed cartesian positions, and a two layer deep (104) × (01$\bar{4}$) calcium-terminated step was used. Calcium-terminated steps were used hence the crystallizations were performed in slow diffusion technique and therefore in excess of calcium in solution. The starting geometries of the additive were produced by the construction of an amorphous cell of 50 molecules for the given size of the bulk crystal. After layering, the geometry was optimized with the Forcite package and the COMPASS forcefield at ultra-fine quality (500 000 iterations at maximum). Afterwards, a 10 ps Forcite quench (ultra-fine, COMPASS forcefield with Ewald summation) was performed to study the stability of surface-bound amino acids.

In circular dichroism experiments, any employed vial was heated out at 110° C for 3 h prior usage. Base solutions of amino acids (alanine, proline, phenyl alanine, valine and as control glycine and RAC-alanine), in case of alanine from different manufactures (Merck, Acros, ABCR, Fluka) and different quality (e.g. 99.98%, Biochem, for synthesis) were produced gravimetrically and were diluted to yield a 2 g l^{-1} amino acid solution. After mixing equal volumes of the respective amino acid base solution and 20 mmol l^{-1} CaCl$_2$, the resulting solution was equilibrated at room temperature for at least 12 h. Any CD measurements (Jasco CDF-426S) were performed three times with at least three different batches of solutions. Data analysis was performed with Origin 7.5 (OriginLab Corp).

References

[1] L. Perez-Garcia and D. B. Amabilino, *Chem Rev* **2002**, *31*, 342.
[2] W. A. Bonner, *Chirality* **2000**, *12*, 114.
[3] M. P. Bernstein, J. P. Dworkin, S. A. Sandford, G. Cooper and L. J. Allamandola, *Nature* **2002**, *416*, 401.
[4] D. Y. Sumner, *Am J Sci* **1997**, *297*, 455.
[5] D. P. E. Smith, *J Vac Sci Technol B* **1991**, *9*, 1119.
[6] S. J. Sowerby, W. M. Heckl and G. B. Petersen, *J Mol Evol* **1996**, *43*, 419.
[7] J. V. Barth, J. Weckesser, G. Trimarchi, M. Vladimirova, A. D. Vita, C. Cai, H. Brune, P. Günter and K. Kern, *J Am Chem Soc* **2002**, *124*, 7991.
[8] H. Spillmann, A. Dmitriev, N. Lin, P. Messina, J. V. Barth and K. Kern, *J Am Chem Soc* **2003**, *125*, 10725.
[9] C. B. France and B. Parkinson, *J Am Chem Soc* **2003**, *125*, 12712.
[10] J. R. Cronin and S. Pizzarello, *Science* **1997**, *275*, 951.
[11] S. Pizzarello and J. R. Cronin, *Geochim Cosmochim Acta* **2000**, *64*, 329.
[12] J. Bailey, A. Chrysostomou, J. H. Hough, T. M. Gledhill, A. McCall, S. Clark, F. Menard and M. Tamura, *Science* **1998**, *281*, 672.
[13] R. M. Hazen, T. R. Filley and G. A. Goodfriend, *Proc Natl Acad Sci USA* **2001**, *98*, 5487.
[14] M. O. Lorenzo, S. Haq, T. Bertrams, P. Murray, R. Raval and C. J. Baddeley, *J Phys Chem B* **1999**, *103*, 10661.
[15] M. Lorenzo, C. J. Baddeley, C. Muryn and R. Raval, *Nature* **2000**, *404*, 376.
[16] R. Humblot, S. M. Barlow and R. Raval, *Prog Surf Sci* **2004**, *76*, 1.
[17] S. Romer, B. Behzadi, R. Fasel and K.-H. Ernst, *Chem Eur J* **2005**, *11*, 4149.
[18] A. Kühnle, T. R. Linderoth, B. Hammer and F. Besenbacher, *Nature* **2002**, *415*, 891.

[19] Q. Chen and R. V. Richardson, *Nat Mater* **2003**, *2*, 324.
[20] C. A. Orme, A. Noy, A. Wierzbicki, M. T. McBride, M. Grantham, H. H. Teng, P. M. Dove and J. J. DeYoreo, *Nature* **2001**, *411*, 775.
[21] H. Meyer, *Angew Chem* **1959**, *21*, 678.
[22] S. Kamhi, *Acta Cryst* **1963**, *16*, 770.
[23] H. Meyer, *Z Kristallographie* **1969**, *128*, 183.
[24] F. Lippmann, *Sedimentary Carbonate Minerals*, Springer-Verlag, Berlin **1973**.
[25] M. Lahav and L. Leiserowitz, *Angew Chem Int Ed* **2007**, *46* (20), 5618.
[26] R. M. Hazen, private communication **2008**.
[27] L. Addadi and S. Weiner, *Proc Natl Acad Sci USA* **1985**, *82*, 4110.
[28] S. Mann, *Struct Bonding* **1983**, *54*, 125.
[29] H. A. Löwenstam and S. Weiner, *On Biomineralization*, Oxford University Press, New York **1989**.
[30] S. Weiner and L. Addadi, *J Mater Chem* **1997**, *7*, 689.
[31] S. Mann, *Biomineralization*, Oxford University Press, Oxford **2001**, ISBN 0-19-850882-4.
[32] F. Meldrum, *Int Mater Rev* **2003**, *48*, 187.
[33] W. Tremel, J. Küther, M. Balz, N. Loges and S. E. Wolf, *Biomineralization*, Wiley–VCH, Weinheim **2007**.
[34] A. Berman, L. Addadi and S. Weiner, *Nature* **1988**, *331*, 546.
[35] G. Falini, S. Albeck, S. Weiner and L. Addadi, *Science* **1996**, *271*, 67.
[36] N. Nassif, N. Pinna, N. Gehrke, M. Antonietti, C. Jäger and H. Cölfen, *Proc Natl Acad Sci USA* **2005**, *102*, 12653.
[37] I. Weißbuch, L. Addadi, M. Lahav and L. Leiserowitz, *Science* **1991**, *253*, 637.
[38] D. Jaquemain, S. G. Wolf, F. Leveiller, M. Deutsch, K. Kjaer, J. Als-Nielsen, M. Lahav and L. Leiserowitz, *Angew Chem Int Ed* **1992**, *31*, 130.
[39] I. Weißbuch, R. Popovitz-Biro, M. Lahav and L. Leiserowitz, *Acta Crystallogr Sect B* **1995**, *51*, 115.
[40] S. Mann, B. R. Heywood, S. Rajam and J. D. Birchall, *Nature* **1988**, *334*, 692.
[41] S. Mann, D. D. Archibald, J. M. Didymus, T. Doughlus, B. R. Heywood, F. C. Meldrum and M. J. Reeves, *Science* **1993**, *261*, 1286.
[42] J. J. M. Donners, R. J. M. Nolte and N. A. J. M. Sommerdijk, *J Am Chem Soc* **2002**, *124*, 9700.
[43] A. M. Travaille, L. Kaptijn, P. Verwer, B. Huisgen, J. A. A. Elemans, R. J. M. Nolte and H. van Kempten, *J Am Chem Soc* **2003**, *125*, 11571.

References

[44] S. Cavalli, D. C. Popescu, E. E. Tellers, M. R. J. Vos, B. C. Pichon, M. Overhead, H. Rapaport, N. A. J. M. Somerdijk and A. Kros, *Angew Chem Int Ed* **2006**, *45*, 739.

[45] T. N. Thomas, T. A. Land, J. J. D. Yoreo and W. H. Casey, *Langmuir* **2004**, *20*, 7643.

[46] K. J. Davis, P. M. Dove, L. E.Wasylenki and J. J. D. Yoreo, *Am Mineral* **2004**, *89*, 714.

[47] K. J. Davis, P. M. Dove and J. J. D. Yoreo, *Science* **2000**, *290*, 1134.

[48] W. Hou and Q. Feng, *J Cryst Growth* **2005**, *282*, 214.

[49] C. Shivkumara, P. Singh, A. Gupta and M. S. Hegde, *Mater Res Bull* **2006**, *41*, 1455.

[50] W. Ostwald, *Z Phys Chem* **1897**, *22*, 289.

[51] Match, Crystal Impact **2005**, database: PDF–2, Release 2004, JCPDS – International Centre for Diffraction Data, Newtown Square, PA 19073–3273, USA.

[52] N. Loges, S. Wolf and W. Tremel, Edging effekt induced by monocarboxylic acids in the growth of calcium carbonate, in preparation.

[53] Materials studio, Accelrys Inc. **2006**.

[54] R. T. Downs and R. M. Hazen, *J Mol Catal A* **2004**, *216*, 273.

[55] V. A. Avetisov, V. I. Goldanskii and V. V. Kuzmin, *Phys Today* **1991**, *44*, 33.

[56] G. Ertem and J. P. Ferris, *J Am Chem Soc* **1997**, *119*, 7197.

[57] Buhse, Durand, Kondepudi, Laudadio and Spilker, *Phys Rev Lett* **2000**, *84* (19), 4405.

[58] S. Veintemillas-Verdaguer, S. O. Esteban and M. A. Herrero, *J Cryst Growth* **2007**, *303*, 562.

[59] C. Viedma, *Cryst Growth Des* **2007**, *7*, 553.

[60] H. Schmidbaur, H.-G. Classen and J. Helbig, *Angew Chemie* **1990**, *102*, 1122.

[61] N. Loges, S. E. Wolf, M. Panthöfer, L. Müller, M.-C. Reinnig, T. Hoffmann and W. Tremel, *Angew Chem Int Ed* **2008**, *120* (20), 3741.

[62] J. H. E. Cartwright, O. Piro and I. Tuval, *Phys Rev Lett* **2007**, *98* (16), 165501.

[63] L. D. Barron, *New Developments in Molecular Chirality*, Kluwer, Dordrecht, The Netherlands **1995**.

[64] L. Keszthelyi, *J Biol Phys* **1994**, *20*, 241.

[65] A. Bakasov, H. Tae-Kyu and M. Quack, *Chemical Evolution: Physics and the Origin and Evolution of Life*, Kluwer, The Netherlands, ISBN 978-079234-111-6 **1996**, 287–296.

[66] P. Lazzeretti and R. Zanasi, *Chem Phys Letters* **1997**, *279*, 349.

[67] A. Szabo-Nagy and L. Keszthelyi, *Proc Natl Acad Sci USA* **1999**, *96*, 4252.

[68] Y. Scolnik, I. Portnaya, U. Cogan, S. Tal, R. Haimovitz, M. Fridkin, A. C. Elitzur, D. W. Deamer and M. Shinitzky, *Phys Chem Chem Phys* **2006**, *8* (3), 333.

[69] M. Shinitzky, A. C. Elitzur and D. W. Deamer, *Progress in Biological Chirality*, Elsevier, New York, ISBN 978-0-08-044396-6 **2004**.

[70] D. W. Deamer and M. Shinitzky, *Astrobiology* **2006**, in press.

[71] J. Küther and W. Tremel, *Chem Commun* **1997**, 2029.

[72] J. Küther, R. Seshadri, W. Knoll and W. Tremel, *J Mater Chem* **1998**, *8*, 641.

[73] J. Küther, G. Nelles, R. Seshadri, M. Schaub, H.-J. Butt and W. Tremel, *Chem Eur J* **1998**, *4*, 1834.

[74] J. Küther, R. Seshadri and W. Tremel, *Angew Chem Int Ed* **1998**, *37*, 3044.

[75] J. Küther, R. Seshadri, G. Nelles, H.-J. Butt, W. Knoll and W. Tremel, *Adv Mater* **1998**, *10*, 401.

[76] M. Balz, H. A. Therese, J. Li, J. S. Gutmann, M. Kappl, L. Nasdala, W. Hofmeister, H.-J. Butt and W. Tremel, *Adv Funct Mater* **2005**, *15*, 683.

[77] M. Balz, H. A. Therese, M. Kappl, L. Nasdala, W. Hofmeister, H.-J. Butt and W. Tremel, *Langmuir* **2005**, *21*, 3981.

[78] Y. Kitano, *Bull Chem Soc Japan* **1962**, *35* (12), 1973.

[79] M. Page and H. Cölfen, *Cryst Growth Des* **2006**, *6* (8), 1915.

References

8 Résumé

Nonclassical crystallization, *i.e.* crystallization processes which challenge classical theories of nucleation and crystal growth, appears today to bear the potential of a toolbox for gaining control over a wide range of material characteristics. Different classes of nonclassical crystallization processes were presented in the preceding chapters: *(i)* liquid amorphous intermediates, *(ii)* mesocrystallization, and *(iii)* symmetry-breaking (*cf.* Research Objectives, Chp. 2). All three classes were shown to control key characteristics of materials like phase, shape or size. As a model system, calcium carbonate was chosen, which on the one hand appears quite simple and is facile to produce but on the other hand it is of remarkable industrial impact and of abundant presence exhibiting delicate molding and extraordinary functionality in Nature.

The phase selection by means of symmetry-breaking interaction of amino acid additives demonstrates a unusual category of nonclassical crystallizations. Symmetry-breaking is of high scientific interest but highly debated. The discovery of the parity violating weak interaction jeopardized foundations of classic chemistry: the principle of symmetry and the enthalpic equality of stereoisomers. The findings, which are presented in Chp. 7, can be regarded as one of today numerous examples of symmetry-breaking by amplification of the parity violating weak interaction by crystallization or polymerization. In the present case, only the levorotatory additives lead to a *nonclassical phase selection* of a metastable calcium carbonate polymorph, which is the first demonstration that the phase selection of $CaCO_3$ can depend on the chirality of the additives.

Nature molds organic components of hard tissue by occlusion in grain boundaries of the mineralized constituents, which yields in the case of the gastropoda *Haliotis laevigata* in a "stack of coins"-like structure. The crystallographically oriented calcium carbonate tablets are separated by small sheets of organic tissue. This concept of molding of organic tissue was adopted in a reversed manner in order to yield in a *molded calcium carbonate hollow structure* (*cf.* Chp. 6). Replacing the soft tissue by calcium carbonate and the hard tissue by a mesocrystalline ice matrix, small sheets of calcium carbonate were obtained. After removing the ice mesocrystal by freeze-drying, these sheets reeled to form nanotubes featuring a remarkable aspect-ratio.

(iii) Nonclassical and Symmetry-Breaking Phase Selection of Calcium Carbonate.

(ii) Nonclassical Compression-Molding of Calcium Carbonate in Ice Mesocrystals.

8. Résumé

Built up from crystallographic aligned nanocrystals, mineral mesocrystals show the potential of adopting different shapes whereas in the presented case, the mesocrystallinity of ice was employed to mold a mineral in a nonclassical shape.

(i) Nonclassical Liquid Intermediates of Bivalent Carbonates... Form Due to Numerous Constituents and Solvent Interaction...

Faatz *et al.* postulated and Rieger *et al.* evidenced the existence of a liquid calcium carbonate intermediate but the reported experiments may suffer from experimental artifacts of fast-mixing which falsely hint at the existence of a calcium carbonate liquid intermediate phase. By a diffusion-controlled and contract-free experimental setup, the existence of *nonclassical liquid intermediates*, which precede the crystalline mineral phase of a bivalent metal carbonate at neutral pH, was proven in Chp. 3 and Chp. 4. The formation and the stability of this extraordinary liquid-like carbonate phase is attributed to the presence of numerous species of carbonate and metal complexes during the precipitation process. A vast variety of carbonato-, bicarbonato- and aquo-coordinated metal complexes which further can condensate to form a polynuclear network consisting of water, metal complexes and the three carbonate species. The presence of various bonding partners, variable coordination geometries and coordination numbers of the carbonate groups and the associated distribution of local structures favors the formation of a non-crystalline phase. In other words, the pH dependent protonation of carbonate, which yields in the two additional species of bicarbonate and carbonic acid, jams the formation of a crystalline phase by lowering carbonate activity and 'disguising' carbonate ions as bicarbonate and carbonic acid. In the present case, the solvent water is able to incorporate and massively interact with the blocked carbonate species and thus leads to a highly hydrated and liquid-like calcium carbonate phase. The lifespan of the emulsified state is remarkable under contact-free conditions; no extensive aggregation and coalescence does occur. The stabilization of the

are Electrostatically Stabilized...

emulsion is electrostatically, its coalescence and aggregation barrier could be easily lowered by the addition of salt which screened the coulomb interaction as predicted by the DLVO theory. In several publications, L. A. Gower *et al.* proposed and evidenced the existence of a so-called 'polymer-induced liquid-precursor' (PILP). Their investigation mainly focused on the late state of the precursor, its transformation to crystalline material and the application of this precursor. However, very little was known about the origin, the mechanism of formation and stabilization of this precursor. As the name implies, it is proposed that the liquid precursor is induced by the addition of tiny amounts of polymers, which gather and sequester cations and as a consequence to attenuate the supersaturation and to delay the crystallization. Proteins of different isoelectric points were employed in Chp. 5, to gain insight into mechanistic details. The basic and positively charged protein lysozyme (pI = 9.3) demulsifies the transient state, so that the assumption of negative surface charge and an electrostatic stabilization of the pure emulsified liquid calcium carbonate precursor is reasonable. This backs the conclusions, which were drawn from the aggregation and coalescence of the emulsified and additive-free state in presence of salt. In contrary to lysozyme, the negatively charged acidic protein ovalbumin ($pI = 4.7$) extends the life-span of the emulsified state. Deduced from recent small-angle neutron-scattering data, ovalbumin dramatically decreases calcium activity of the bulk solution but locally increases the calcium concentration next to the protein. The decrease in calcium activity relieves the solution's supersaturation, prevents the formation of crystalline material, whose formation would require higher supersaturation according to their solubility products, and in summary extends the life-time of the liquid and amorphous precursors. From a colloidal chemical point of view, this was interpreted as a depletion stabilization in Chp. 5. As the process of depletion stabilization is invariably preceded by depletion flocculation, the PILP coating effect actually evolves to be a result of a demulsifying process induced by depletion flocculation.

and Comply with the DLVO Theory.

In brief, three categories of nonclassical crystallization of bivalent metal carbonates were investigated: *(iii)* symmetry-breaking phase-selection, *(ii)* mesocrystallization, and *(i)* the emergence of liquid amorphous intermediates. The latter were investigated applying an ultrasonic levitation technique in order to gain insight in the formation and stabilization mechanisms, which have to be described in terms of the physical chemistry of colloids and emulsions. This findings may pave the way for a broader application of the PILP mechanism and for a deeper understanding of biomineralization processes *in vivo*.

8. Résumé

Appendix

Classical Concepts of Crystallization

Phase transitions cannot be described based on the so-called thermostatics (the thermodynamics of equilibrium processes) since they are irreversible. The supersaturation S is a quantity which allows one to describe in simple terms how far away a system is from a state of equilibrium. It is defined as a ratio of the actual concentration of the solute c to the equilibrium solubility product K_{SP}.

Supersaturation

$$S = \frac{c}{K_{SP}} \qquad \text{(Eq. A.1)}$$

The change in the associated chemical potential $\Delta\mu$ forces back the system to the equilibrium, which implies that the system will lower its supersaturation via phase separation in general or nucleation in special.

$$\Delta\mu = -kT \ln S \qquad \text{(Eq. A.2)}$$

Gibbs derived in his classical treatments of heterogenous equilibrium, that a necessary condition for stability resp. metastability of a fluid phase is that the chemical potential of a component has to increase with increasing concentration of that component.[1] In case of two components this reduces to $(\partial^2 G/\partial c^2)_{T,P} > 0$. Thus the limit of metastability is $(\partial^2 G/\partial c^2)_{T,P} = 0$, which is called the spinodal line. A phase diagram of a regular solution is depicted in A.3; the binodal line divides the regions of stability and metastability, whereas the spinodal line separates the regions of meta- and instability. In the metastable region, a finite concentration fluctuation—the nucleus— is needed to destabilize the solution. The energy barrier of nucleation is a measure of the solution's metastability. At the spinodal line, the work of nucleation approaches zero which leads to joint concentration fluctuations which spontaneously grow. We will first investigate binodal processes, nucleation, and then turn to spinodal decompositions, which are described by the Cahn-Hilliard theory.

Conditions of Stability

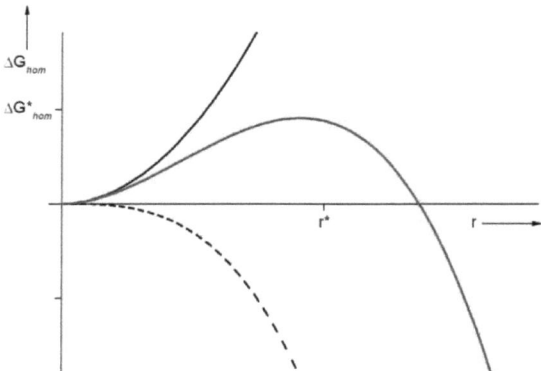

Figure A.1 The molar free enthalpy during the growth of a nucleus. The red line represents $\Delta G_{hom}(r)$, which is the sum of the stabilizing volume term ΔG_V (black line) and the destabilizing surface term ΔG_S (dashed black line).

Phase Separation in the Metastable Region

Classical Nucleation Theory   The classical approach of Becker, Döring and Volmer describes nucleation in terms of an aggregation of single building blocks (molecules or ions).[2,3] The driving force of homogenous nucleation arise from the change ΔG_{hom} in free enthalpy, which is the sum of a stabilizing volume term ΔG_V and a destabilizing surface term ΔG_S. The volume term ΔG_V quantifies the gain in molar free enthalpy $\Delta G_N(n)$ if n building blocks are incorporated in the nucleus (thus in the new phase). If the number of building blocks n is expressed in terms of nucleus volume V_N and molar volume V_M of the building blocks, then the volume term will scale linearly with the nucleus volume V_N. The loss in free enthalpy ΔG_S, which is needed to create the new interphase A_N (thus the surface of the nucleus), is a function of the surface free energy γ. Generally, the surface free energy γ is assumed to be independent of the curvature of the nucleus, primarily because $\gamma(r)$ cannot be determined experimentally merely molecular-dynamically. The Laplace formula predict a higher pressure due to the surface tension. Assuming an incompressible nucleus, this effect is commonly neglected.

$$\Delta G_{hom} = \Delta G_V + \Delta G_S = \frac{\Delta G_N(n)}{V_M} V_N - \gamma \cdot A_N \quad \text{(Eq. A.3)}$$

$$= \frac{4\pi \rho_N}{3 M_N} \Delta G_N(n) r^3 - 4\pi \sigma_N r^2 \quad \text{(Eq. A.4)}$$

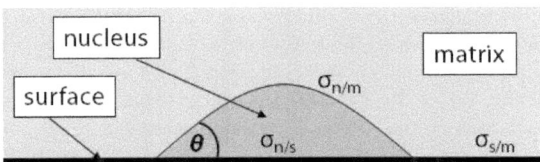

Figure A.2 Principle of an heterogenous nucleation. Each interphase is characterized by its surface tension σ.

Assuming that the nuclei are spherical, one can rearrange the equation Eq. A.3 to yield equation Eq. A.4 which is plotted in A.1. The curve's angular point corresponds to the nucleation barrier $\Delta G^*_{hom} = \Delta G_{hom}(r^*)$ which a nucleus has to overcome in order to represent a new stable phase. Thus, nucleus has to reach a critical radius $r^* = 2\sigma_N M_N / \rho_N \Delta G_N(n)$ to withstand redissolution.

The classical homogenous nucleation rate J, which is a product of the nucleus number density and the diffusion current, can be approximated by a Boltzmann approach.

$$J_N \propto e^{-\Delta G^*_{hom}/k_B T} \quad \text{(Eq. A.5)}$$

If foreign bodies affect the nucleation process (e.g. vessel walls, proteins and their aggregates, dust, ...), heterogenous nucleation may occur. A heterogenous nucleation as depicted in Fig. A.2 is energetically favored because the surface term ΔG_S in Eq. A.3 is reduced by a factor $f(\theta)$; Eq. A.6 gives the example of $f(\theta)$ in case of a flat heterogenous interface. Hence, it is difficult to study pure homogenous nucleation and to suppress any heterogenous influences. Only if the contact angle θ approaches the upper limit of 2π no heterogenous influence would occur. This probably will never occurs in reality. The contact angle is defined in Young's equation (Eq. A.7), which describes the equilibrium contact angle of a droplet placed on a flat surface.

Heterogenous Nucleation

$$f(\theta) = \frac{1}{4}(2 + \cos\theta)(1 - \cos\theta)^2 \quad \text{(Eq. A.6)}$$

$$\cos\theta = \frac{\sigma_{s/m} - \sigma_{n/m}}{\sigma_{n/s}} \quad \text{(Eq. A.7)}$$

Phase Separation in the Instable Region

Cahn and Hilliard studied phase separation in the unstable region which occurs by a spinodal mechanism. A complete derivation is skipped in favor

Spinodal Phase Separation

of a short discussion of the findings concerning crystallization; for a more elaborate derivation please refer to the corresponding literature.[4] If a system of two components is quickly destabilized (e.g. by decreasing temperature quickly enough, Fig. A.3), it reaches the region of instability. Now, smallest concentration fluctuations are no more damped out but they spontaneously grow (Fig. A.4a–c). In other words, the diffusion coefficient of the solute component is negative and the diffusion is directed uphill towards the concentration gradient. Following the Cahn-Hilliard theory, the fluctuations are spatially related by an appropriate characteristic and time-dependent wavelength $\lambda(t)$ (Fig. A.4a). Finally, the concentration fluctuations are static and the phase boundaries are stable. The system has reached a two-phased state of equilibrium as it has left the in- and metastable regions (Fig. A.3 and A.4d). The kinetic of such a process is described by the Cahn-Hilliard equation (Eq. A.8). It is highly non-linear because the chemical potential μ is a function of the concentration c. Patterns, which occur during and because of spinodal phase separations, can be modeled applying this equation.

$$\dot{c} = D_0 \nabla^2 \mu \qquad\qquad \text{(Eq. A.8)}$$

Further Crystal Growth

The formation of a stable nucleus is a stochastically seldom event. If this occurs, the nucleus starts its growth. The classical model subdivides the incorporation of a single building block in a crystal in several consecutive steps. First, *(a)* the building blocks have to diffuse to the crystal. After their *(b)* absorption on the surface, they undergo *(c)* two-dimensional diffusion on the crystal surface to an active growth step (Volmer diffusion). After reaching the growth step, *(d)* one-dimensional diffusion to an existing kink site may occur before the building block is finally gets *(d)* incorporated in the bulk crystal.

Volmer Diffusion

Kossel-Stranski Model

If one layer is complete, new active growth sides will be crucial for a continuing growth of the crystal. Single building blocks may attach to the surface independently of the further growth of layer, which only occurs at kink sites as this is clearly energetically favored. Only if the actual growing layer is completed, the first attaching building block acts as a nucleus for the next new-forming layer. Thus, in the classical model of Kossel and Stranski, who proposed independently the same model, the crystal grows layer by layer.

Frank Model

However and as so often, the observed rates of crystal growth at low supersaturation are too rapid to be explained by nucleation on flat surfaces. Frank deduced, based on experimental data, the existence of screw dislocations and presented their possible importance on a Faraday Discussion

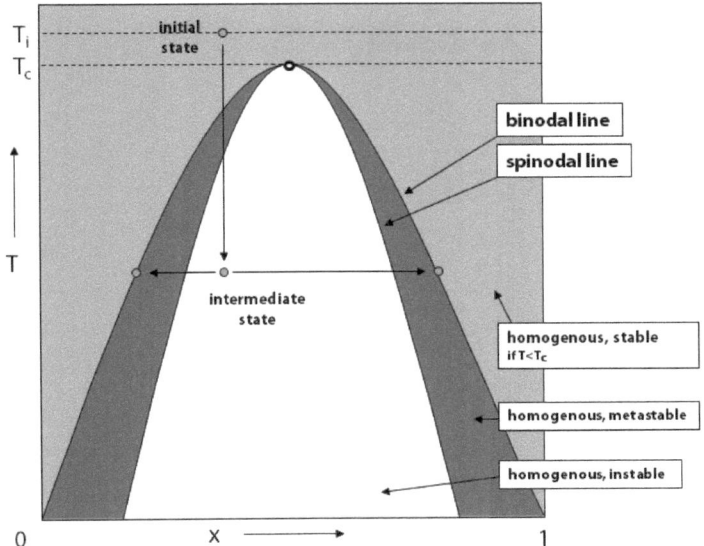

Figure A.3 Idealized phase diagram of a two-component system. Starting from the initial state characterized by $T_i > T_c$, the system is instable and enters the spinodal region so fast that no binodal phase separation occurs, a spinodal phase separation can take place. [4]

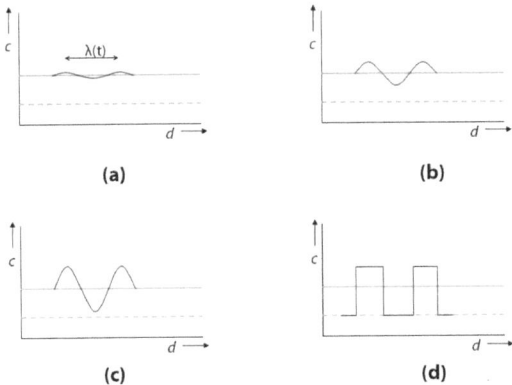

Figure A.4 Course of a spinodal decomposition. **(a–c)** A small fluctuation in concentration is amplified by diffusion. **(d)** Finally, the system reaches a stable two-phased state. [5]

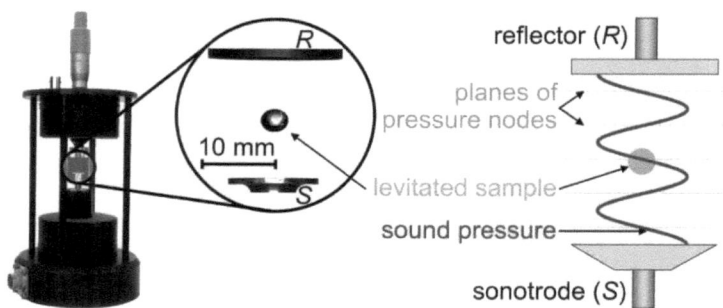

Figure A.5 Setup and principle of the employed acoustic levitator. A sound level of about 160 dB compensates gravity. A droplet of water is levitated in the central node of the standing acoustic wave to demonstrate the levitation of liquid samples.

of Crystal Growth.[6] Shortly afterwards, experimental verification of his assumptions followed.[7-9]

Basic Principles of Acoustical Levitation

First described in 1933 by Bücks and Müller,[10] the acoustic levitation technique has attracted considerable attention until today in various fields, which require a reliable contactless environment. The acoustic levitator, which was employed in this dissertation, is depicted in Fig. A.5. As a result of axial radiation pressure and radial Bernoulli forces, liquid and solid samples can be placed and held in a levitated position in the sound pressure nodes of this wave without contact. Typically, levitated samples have a volume of 5 nL – 5 µL (corresponding to a diameter of 0.2 – 2 mm). No other constraints, such as magnetic or dielectric properties, apply for the ultrasonic leviation of a sample.

A standing acoustic wave can only be generated if the distance l^* between sonotrode S and reflector R is an integer multiple of the half wavelength λ.

$$n \cdot \frac{\lambda}{2} = l^* \qquad n = 1, 2, 3, \ldots \qquad \text{(Eq. A.9)}$$

The levitational force, which has to act on a aqueous droplet of 5 µL can be estimated quite easily. The gravitational force $F_G = mg \approx 50\,\mu\text{N}$ has to be compensated by the sum of levitational force F_L and buoyant force F_B. In case of an aqueous droplet, the latter force is in the negligible range of 65 nN.

Appendix

A mathematical derivation of the levitational force F_L starts from the equation of continuity (Eq. A.10) and the equation of motion of the gas pressure (Eq. A.11).[11]

$$\frac{\partial}{\partial t}\rho + \nabla \rho \underline{q} = 0 \qquad \text{(Eq. A.10)}$$

$$\rho \cdot \mathfrak{D}\underline{q} = -\nabla p \qquad \text{(Eq. A.11)}$$

$$\text{where } \mathfrak{D} = \frac{\partial}{\partial t} + v_x\frac{\partial}{\partial x} + v_y\frac{\partial}{\partial y} + v_z\frac{\partial}{\partial z}$$

Herein, the vector $\underline{q} = f(v_x, v_y, v_z)$ describes the velocity of gas particles as a function of location. Now, several approximations are introduced. The density of the medium ρ is local invariable and the absence of turbulence near the levitated particle is assumed. Along with boundary conditions concerning the integral of pressure variance at the levitated particle surface, the levitational force F_L for a droplet resp. sphere can be expressed with the following and concluding equation.

$$F_L = \frac{5\pi^2}{3}\frac{r^3}{\lambda}\rho_E \sin\left(\frac{4\pi}{\lambda}\Delta x\right) \qquad \text{(Eq. A.12)}$$

Avrami's Equation

In the 1930s, Avrami developed a theory of the kinetics of phase change, yielding the so-called Avrami equation.*,[12]

$$f_V = 1 - e^{-k_A t^n} \qquad \text{(Eq. A.13)}$$

As the fractional volume f_V of the new phase is normalized to run betwen zero and one, f_V is equatable with scale factors $s(t)$ which can be extracted by a Rietveld analysis[13] of X-ray powder data. The exponent n is called the Avrami exponent, which mainly characterizes the crystallization kinetics and k_A is the Avrami constant. Hoewever, the physical interpration of these both constants remains difficult.[14,15]

Presented by Jena et al.,[14] a simple derivation of the Avrami equation analyzes a transformation of a phase α to a new phase β occuring in the time interval $t_0 < t < t_{max}$ and starts with the following assumptions: **(i)** In

---

* Also known as the Kolmogorov-Johnson-Mehl-Avrami (KJMA) equation

untransformed material α, the nucleation of β occurs unhindered and randomly. (ii) The nucleation rate per volume $\dot{\mathcal{N}}$ of β remains constant over the complete transformation process. (iii) The growth rate $\dot{\mathcal{G}}$ of the nucleated β particles is isotropic.

During a time interval of dt, $\mathcal{N}$ new β particles will form in the untransformed α phase.

$$\mathcal{N} = \dot{\mathcal{N}} V dt$$

In the remaining time of the transformation process $(t_{max} - t)$, the formed nuclei will grow to yield a spherical particle of radius $\dot{\mathcal{G}}(t_{max} - t)$. If we initially neglect the loss in volume of the α phase, the phase β would increase and the so-called 'extended volume' $dV\beta_{ext}$ of the time interval dt would be Eq. A.14 and its integral V^{β}_{ext} determines the total extended volume in the time interval $(t - t_{max})$ (Eq. A.15).

$$dV^{\beta}_{ext} = \frac{4\pi}{3} \left(\dot{\mathcal{G}}(t_{max} - t)\right)^3 \dot{\mathcal{N}} dt \qquad \text{(Eq. A.14)}$$

$$V^{\beta}_{ext} = \frac{\pi}{3} V \dot{\mathcal{N}} \dot{\mathcal{G}}^3 t^4 \qquad \text{(Eq. A.15)}$$

As a portion of the extended volume "overlaps" with already transformed volume, the volume of really new formed β is proportional to the volume fraction of α.

$$dV^{\beta} = V^{\beta}_{ext}\left(V - V^{\beta}\right) \qquad \text{(Eq. A.16)}$$

Integration and rearrangement yields the familiar form of the Avrami equation.

$$f_V(t) := \frac{V^{\beta}}{V} \qquad \text{(Eq. A.17)}$$

$$= 1 - e^{-k_A t^n} \qquad \text{(Eq. A.18)}$$

$$\text{where } k_A = \frac{\pi}{3}\dot{\mathcal{N}}\dot{\mathcal{G}}^3 \text{ and } n = 4$$

References

[1] J. W. Gibbs, *Collected Works*, Yale University Press, New Haven, Conneticut **1948**.
[2] M. Volmer, *Kinetik der Phasenbildung*, Steinkopff, Dresden **1939**.
[3] R. Becker and W. Döring, *Ann Phys* **1935**, *24*, 719.
[4] S. Komura, *Phase Trans* **1988**, *12* (1), 3.

[5] R. W. Cahn, P. Haasen and E. J. Kramer (eds.), *Phase Transitions of Materials*, vol. 5, VCH-Verlag, Weinheim **1991**.

[6] F. C. Frank, *Adv Phys* **1952**, *91*, 1.

[7] J. L. Griffin, *Phil Mag* **1950**, *41*, 196.

[8] S. Amelinckx, *Nature* **1951**, *167*, 939.

[9] S. Amelinckx, *Nature* **1951**, *168*, 431.

[10] K. Bücks and H. Müller, *Z Physik* **1933**, *84*, 75.

[11] A. Stockhaus, *Aufbau und Entwicklung einer akustischen Falle*, Master's thesis, Institut für Spektrochemie und angewandte Spektroskopie Universität Dortmund **1997**.

[12] A. R. West, *Solid State Chemistry and its Applications*, John Wiley & Sons, Chichester, England **1984**.

[13] H. M. Rietveld, *J Appl Cryst* **1969**, *2*, 65.

[14] A. K. Jena and M. C. Chaturvedi, *Phase Transformations in Materials.*, Prentice Hall **1992**.

[15] Y. Khanna and T. Taylor, *Polym Eng Sci* **1988**, *28*, 1042.

List of the Appendix' Variables and Constants

c_0 speed of sound in air

i imaginary unit

A_N nucleus surface

A_p amplitude of the acoustic field at the particle surface

A_0 amplitude of the incident wave at the source

A_0^* effective amplitude of the acoustic field

c solute concentration

d spacial distance in bulk solution

D_0 diffusion coefficient

F_B buoyant force

F_G gravitational force

F_L levitational force

$f(\theta)$ homogenous nucleation factor

f_V fractional volume in the Avrami equation

$G, \Delta G$ Gibbs' enthalpy and its change

$\Delta G_S, \Delta G_V$ changes in surface enthalpy and volume enthalpy

$\Delta G_N(n)$ molar free enthalpy of a nucleus as a function of the number of incorporates building blocks n

ΔG_{hom}^* nucleation barrier

$\dot{\mathcal{G}}$ Growth rate of a nucleus

J_N homogenous nucleation rate

k_A Avrami constant

k_0 wave number

l axial coordinate between sonotrode S and reflector R

l^* distance between sonotrode S and reflector R

$\lambda(t)$ Cahn-Hilliard length

k_B Boltzmann constant

K_{SP} equilibrium solubility product

λ wavelength of the acoustic standing wave

n number of building blocks incorporated in the nucleus or Avrami exponent

$\mathcal{N}$ nucleation particle number

$\dot{\mathcal{N}}$ nucleation rate per volume

ν_0 velocity of sound

M_N molar mass of building blocks

$\mu, \Delta\mu$ chemical potential and its change

ω angular frequency corresponding to the ultrasonic range

P pressure

p_i incident pressure wave

p_s pressure wave scattered by the levitated sample

r particle radius, e.g. of levitated droplet or nucleus

r^* critical radius of the particle

ρ density of the gas phase

ρ_N molar density of a nucleus' building blocks

ρ_E density of energy

S supersaturation

σ_N surface tension of the nucleus in bulk solution

$\sigma_{x/y}$ surface tension of a interphase (where n: heterogeneously induced nucleus, s: nucleation inducing surface and m matrix resp. bulk solution)

t time

T, T_i, T_c actual temperature, initial temperature, critical temperature

θ contact angle

V_N, V_M volume of the nucleus, volume of one mole of building blocks

V^α, V^β Volume of the phase α resp. β

$dV\beta_{ext}$ extended volume of phase β in the time-frame dt

V^β_{ext} total extended volume

v_x, v_y, v_z velocity of gas along the indexed cartesian axis.

x phase composition

x cartesian coordinate

Δx distance of the center of mass of the levitated particle to the next upper pressure node

y, z cartesian coordinate

List of Figures

1.1	Comparison of geological and biogenic shaping of calcium carbonate reported by Ernst Haeckel and Victor Goldschmidt	3
1.2	Two representatives of calcareous algae *Emiliana huxleyi* and *Discosphaera tubifera*	3
1.3	Scanning electron micrographs of an *Echinoderm* skeleton	6
3.1	Photograph and mode of operation of the acoustic levitator	15
3.2	Scattering curves recorded during the evaporation of a droplet of saturated $Ca(HCO_3)_2$ solution in contact-free environment.	17
3.3	Microscopy images of a levitated droplet and a representative 2D detector frame	18
3.4	Transmission and scanning electron micrographs of calcium carbonate obtained under levitated conditions	19
3.5	Transmission electron micrographs of crystalline calcium carbonate particles obtained in presence of foreign gold nuclei	20
3.6	Transmission electron micrographs of crystalline calcium carbonate particles obtained in presence of foreign $CaCO_3$ nuclei	20
3.7	Evolution of the profile of the (104) reflection with increasing sample concentration	21
3.8	Cryo-SEM studies on a droplet levitated for 400 s	21
3.9	Time evolution of normalized scale factors $s(t)$	23
3.10	TEM micrographs of the liquid calcium carbonate prepared in presence of NaCl suffering from severe aggregation and coalescence	25
3.11	Experimental setup at the µSpot beamline	27
3.12	Geometric derivation of sample diameter	28
3.13	Geometric derivation of crystallite size	29
4.1	Transmission electron micrographs of liquid-like particles formed by divalent metal carbonates	36
4.2	Wide-angle scattering monitored during evaporation of a saturated divalent barium and cadmium bicarbonate solution under levitated contact-free conditions	37
4.3	Precursor particles of the liquid-like barium carbonate mineral phase which experienced a forced flow	38

List of Figures

5.1 Evolution of scattering intensities during *in situ* monitoring of the evaporation of a levitated calcium bicarbonate solution in presence and absence of proteins 45

5.2 Integral Intensity of the (104) reflexes as a function of time and WAXS pattern of the final stages of crystallization . . . 46

5.3 Scanning electron micrographs of the final stages of precipitation of pure calcium carbonate and in presence of ovalbumin 47

5.4 Standard and transmission electron micrographs of ovalbumin fibrils 47

5.5 Transmission electron micrographs of precipitations of pure calcium carbonate and in presence of lysozyme and ovalbumin 49

5.6 Transmission electron micrographs of immunogold labeled samples precipitated in presence of ovalbumin 50

6.1 Scanning electron micrographs of a frozen droplet of saturated $Ca(HCO_3)_2$ before and after freeze-drying 59

6.2 Scanning electron micrographs of intersticial deposited solute and suspensate showing grain boundaries of a frozen droplet 60

6.3 Transmission electron micrographs showing the incipient reeling of a sMOF sheet 60

6.4 Transmission electron micrographs of $CaCO_3$ and sMOF nanotubes 61

6.5 Wide-angle X-ray scattering of frozen and molten droplets containing saturated calcium bicarbonate solution or pure water 61

6.6 Scheme of the employed cryo-molding of nanotubes 62

7.1 Scanning electron micrographs demonstrating the phase selection during crystallization of calcium carbonate in presence of chiral amino acids 69

7.2 Scanning electron micrographs of crystallizations in presence of achiral and sterically hindered amino acids 70

7.3 Progression of pH and $[Ca^{2+}]$ during a crystallization of $CaCO_3$ in presence of L- and D-alanine 71

7.4 Diffraction patterns of calcium carbonate obtained under addition of L-, D- and racemic alanine 73

7.5 Comparison of binding energies of monofunctional additives and bidentate ampholytes 73

7.6 Confocal laser scanning microscopy images of $CaCO_3$ grown in the presence of fluoresceine 74

7.7 Molecularly modeled binding motif of a racemic mixture of alanine 74

7.8 Detailed view of the calculated binding motif of L- and D-alanine attached to (104) growth steps 75

7.9	Selection of octahedral calcium-alaninato coordination compounds	77
7.10	CD spectra of proline and alanine under addition and absence of Ca^{2+}	81
7.11	Scanning electron micrographs of crystallizations under addition of L-amino acids in water and in 80% deuterated water	82
A.1	The molar free enthalpy during the growth of a nucleus . . .	94
A.2	Principle of an heterogenous nucleation	95
A.3	Idealized phase diagram of a two-component system	97
A.4	Course of a spinodal decomposition	97
A.5	Setup and principle of the employed acoustic levitator	98

I want morebooks!

Buy your books fast and straightforward online - at one of world's fastest growing online book stores! Environmentally sound due to Print-on-Demand technologies.

Buy your books online at
www.morebooks.shop

Kaufen Sie Ihre Bücher schnell und unkompliziert online – auf einer der am schnellsten wachsenden Buchhandelsplattformen weltweit! Dank Print-On-Demand umwelt- und ressourcenschonend produziert.

Bücher schneller online kaufen
www.morebooks.shop

KS OmniScriptum Publishing
Brivibas gatve 197
LV-1039 Riga, Latvia
Telefax: +371 686 204 55

info@omniscriptum.com
www.omniscriptum.com

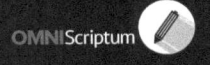

Printed by Books on Demand GmbH, Norderstedt / Germany